Thirteen Years of Travel and Exploration in Alaska

Northern History Library

Thirteen Years of Travel and Exploration in Alaska

1877-1889

W. H. PIERCE

Edited and updated by
R. N. De Armond

ALASKA NORTHWEST PUBLISHING COMPANY
Anchorage, Alaska

Thirteen Years of Travel and Exploration in Alaska,
by W.H. Pierce (edited by Professor and Mrs. J.H. Carruth),
was originally published in 1890 by the
Journal Publishing Company, Lawrence, Kansas.

Library of Congress cataloging in publication data:
Pierce, W H
Thirteen years of travel and exploration in Alaska.
(Northern history library)
1. Alaska—Description and travel—1867-1896.
2. Alaska—Gold discoveries. 3. Gold mines and mining—Alaska. 4. Pierce, W. H. I. Title.
F908.P59 1977 917.98'04'3 77-2893
ISBN 0-88240-076-2

Design by Hilber P. Nelson
Alaska Northwest Publishing Company
Box 4-EEE, Anchorage, Alaska 99509

Printed in U.S.A.

Contents

Editor's Note

Walter H. Pierce was one of the real pioneer prospectors and miners of Alaska. Starting in 1874, there was a fairly sizeable gold rush to the Cassiar District of northern British Columbia. Hundreds of men took part, but few of them stayed in the North. Pierce, who came north in 1877, was one of the relatively small number who did stay on, moving to the Alaska side of the boundary and trying their luck at Windham Bay or Sitka or Juneau or, eventually, the Yukon. As he relates in his book, Pierce tested the ground in all four places.

Except for what he set down in his book, little is known about Walter Pierce. He speaks of himself as a young man in 1877, and when Navy officers took a census at Sitka early in 1880, he was listed as single, an American citizen, and a miner.

Pierce was one of the first to arrive on Gastineau Channel, at what is now Juneau, following the gold discovery by Joe Juneau and Richard Harris. They located their discovery claims on October 4, 1880. On November 2 Pierce, in association with John Dix, Hugh Campbell, O.M. Cole and William Hicks, staked five placer claims "situated below the first canyon of Gold Creek." On February 9, 1881, Pierce attended the miners meeting which organized the mining district, adopted regulations for staking claims, and chose the name

Rockwell for the camp that later became Juneau. In March he staked two lots in the downtown area but in May sold them to Mrs. George Pilz.

As did most of the miners who came to Gastineau Channel, Pierce staked many claims, both placer and lode. Most of them turned out to be worthless, but he actively mined some of the placers and presumably made at least a living. He was associated with John Dix and Hugh Campbell, both experienced miners and both active for many years in the Juneau mining area. Late in 1881 and early in 1882, Pierce and Dix built a water ditch on Specimen Gulch, a tributary of Gold Creek, to carry water for sluicing their diggings. Later in 1882 Pierce and others staked a number of placer claims on Douglas Island "opposite the mouth of Gold Creek." In 1884 he was foreman of the Specimen Gulch Placer Mining Company and seems to have been associated with Dix and Campbell in that operation.

Most of the main events in Pierce's narrative can be documented from other sources. The exception appears in Chapter XXV. This writer, at least, has found nothing to substantiate the story told in that chapter.

And there was one important event that Pierce completely failed to record in his book. This was a murder charge against him, his trial and acquittal.

Details are given in the Epilogue. The illustrations of Fort Wrangel, Sitka, the Stickeen [Stikine] Delta, the Indian dugout canoes, Tlingit dance and the placer miners have been added for historical interest; the line drawings of the totem poles, trading with cannibals and the bear encounter appeared in the original edition. Chapter XXXII, "Some Clippings from Late Alaska Papers," has been omitted as it did not relate to Pierce's story. Otherwise, this edition follows the text of the original, with some clarifying or supplementary information either bracketed in the text or in footnotes.

—R. N. De Armond

Indian totem poles.

FROM THE ORIGINAL EDITION

Introduction

IT IS THE INTENTION OF THE WRITER of this book to put before the reader the knowledge he has gained of the Territory of Alaska, both of the coast and interior, its gold and silver mines, its fisheries; in fact, to give a truthful and reliable account of all its valuable resources and geological and botanical curiosities. A true and reliable knowledge of these things has been obtained in thirteen years of travel and exploration in this little known Territory.

This Territory is larger than that of the thirteen original States of the Union, while its coast line, including bays and islands, is more than 25,000 miles; but it is less known in its interior than Australia or Africa.

It is a country rich in the precious metals, furs, fisheries, and other valuable products.

Some accounts have been published prior to this, generally describing the coast, and what could be seen of this strange land from the deck of a steamship.

These accounts have been mingled with fiction and conclusions drawn from hasty observations. They have not been of a character to truthfully enlighten the reader, as they generally treat of the coast and its scenery. Little or nothing has been said by those writers regarding the more interesting and great unknown interior — its great rivers, forests, and boundless uninhabited wastes and solitudes. Much will be found in this book that will benefit those who contemplate a journey to that distant land, whether it be for profit or pleasure.

My first journey was to northern British Columbia; my second was along the coast of Alaska; my third and fourth were in the interior of Alaska.

Chapter I

ON THE FIRST DAY OF MAY, 1877, a party of five young men, myself included, left the mining town of Nainimo [Nanaimo], British Columbia, our destination being the then famous Cassiar gold fields in the northern British possessions on the head waters of the Mackenzie River, which flows northward and empties into the Arctic Ocean. Being young and hopeful, and in the best of spirits, some of the more sanguine almost considered their fortunes made when the journey began. Little did they know of the years of hardship which some of the little band would endure, this being the experience of all prospectors and frontier adventurers.

The voyage by the steamship *Grappler* from Nainimo, British Columbia, to Fort Wrangel, Alaska, required six days. We called at Fort Simpson, an old Hudson's Bay trading post.

The passengers were a motley crowd — miners, Indians and half-breeds, with quite a number of squaws, the wives of the natives on board. Also some of them were the wives of a class of white men who, having almost lost their taste for civilized life, had taken native women for wives. They were trying a mode of life that could hardly be called civilized, yet it was not wholly uncivilized.

Every spring these men would go to the Cassiar mines, taking their dusky mates with them, and it was no uncommon sight to see them assisting their lords and masters in working their claims. But the majority of them were not so inclined. On the contrary, they appeared to be rather expensive in their habits, and I have seen them sitting by their smoky, dusty camp-fires dressed in silk. This was rather unbecoming for people who, instead of being raised in luxury, had been raised on dried fish, fresh fish, and all kinds of fish — and lived in bark houses with an open, smoky fire-place in the center.

There was a good deal of talk on board the steamer about a quantity of gold that had been taken from a place on Queen Charlotte's Island a great many years ago by the Hudson's Bay Company. The sum, my informant stated, amounted to near $100,000.

According to the story, the natives had discovered the gold on the face of a perpendicular cliff which rose from the salt water, and not knowing what the metal was had taken it to the Hudson's Bay Company's employees. They understood its value, and induced the natives to show them the place where they got it.

As stated above, they succeeded in getting about one hundred thousand dollars. When the vein dipped under the sea water, they built a coffer dam, but had to give it up, as they could not keep the water out. I afterwards inquired of an old Hudson's Bay superintendent about this gold deposit. He told me it was true, and that other parties had afterwards tried to work the lode, but did not succeed in finding gold enough to pay expenses.

On the sixth day of our voyage we reached Fort Wrangel, the winter quarters of the miners, or the major portion of those who work in the Cassiar, or Stickeen [Stikine] River diggings. Here we disembarked. We found it a dirty place. The streets, if they could be called streets, were extremely muddy and crowded with dirty natives and miners, waiting for the ice in the Stickeen River to break up. Then the stream would become navigable for the little steamer that every spring carried the miners and the

The Stickeen [Stikine] Delta.

freight to the head of navigation about one hundred and sixty miles up the river from its mouth.

The river having opened we procured our supplies and got aboard. The boat was crowded to such an extent that there was little more than standing room. When night came on, the boat was tied to a large cottonwood tree, and all hands went ashore to find a place on the damp earth to stretch their weary limbs.

The evening of the second day we reached Glenora Landing, the head of navigation, and disembarked. From here the balance of the journey was over a trail built by the government to the head of Dease Lake.

The morning after our arrival at Glenora Landing was spent in making our baggage into portable parcels and dickering with the packers to get them to a creek called Twenty-four Mile Creek, where we wished to spend at least a few days in search of gold before going on to the principal gold fields.

All things being arranged, the next morning the overland journey commenced. The country traversed by the trail is mountainous, and the trail itself hot and dusty — in places passing through swamps miles in extent, made passable for pack animals by corduroy bridging.

Here the romance of our situation began to wane. The mosquitoes were extremely plentiful, and by no means considerate of our welfare. They buzzed in our ears, got into our eyes, noses and mouths; — in short, made it so uncomfortable for us that we were greatly relieved by camping and making smoke fires to drive them away. Those bloodthirsty insect pests are very plentiful all over the northern part of North America, and especially so in the neighborhood of swamps, making some places almost uninhabitable.

The evening of the second day's journey we camped at Twenty-four Mile Creek, where we wished to try our luck. The next morning we shouldered our tools and started up stream, to use a miner's expression, looking for a favorable bar. Having ascended the creek several miles and prospected several bars, we found some gold, but not enough to pay for working.

We then concluded to move on to Dease Creek, but the pack train had left, and there was only one way to do it. That was to pack the goods ourselves — the different articles of bedding

The steamboat wharf and warehouse at Wrangell in the 1870's, with a portion of old Fort Wrangel at the left, behind its stockade. Shown in the left foreground are two Indian dugout canoes and a Stikine River rowboat with its upturned bow. Wrangell at that time was the transfer point for cargo and men going up the Stikine River to Glenora and thence overland to the Cassiar gold diggings around Dease Lake. River traffic was handled both by small steamers and by Indian canoes.

and tools, cooking utensils and provisions, amounting to more than one hundred pounds to the man. This was rather an uncomfortable situation. The sun was very hot, and the mosquitoes so thick that they would almost blind one. We had to pack our goods sixty miles over this rough mountain trail to the head of Dease Lake, where we could get conveyance by boats to Dease Creek. We shouldered our loads and started, joking each other considerably about our different merits and abilities to perform mules' work. But the inclination to joke soon left us as we jogged along under our heavy loads, some swearing at the vicious mosquitoes, others bearing their burdens in silence.

I observed then, as I have noticed on numerous occasions since, that the silent and patient ones far excelled in endurance those who used profane language and openly complained.

After several days of sweating and toiling, we arrived at the head of navigation on the Arctic Slope, where, procuring boats, we loaded on our tools and supplies and proceeded without further trouble to the diggings.

Here our faith in the richness of the diggings was considerably diminished by our observations. The miners all appeared to be poor, and a large majority of them were subsisting on credit, as they remarked, until their dead work was done and they could get their claims properly opened and on a paying basis. Many wanted to sell out, and after a great deal of talk regarding the richness of their claims would offer to sell at a price so low that one would naturally wonder why a man should want to give away thousands for hundreds.

Years afterwards I understood these things better. Provisions in the diggings were high — flour, fifty cents per pound, and all other things in proportion. The country is extremely mountainous, and the mountains are all capped with eternal snow. Timber grows only on the low lands. All transportation of supplies is done in river boats, and where there are no rivers it is done by Indians who charge very high for packing. In those rugged mountains it would be almost impossible to make trails fit for animals.

These mines were discovered in the year 1873, by Henry Tibbetts [Thiebert] and J. McCulloch, two Canadians, who, having the spirit of adventure, crossed the Rocky Mountains from the east side with a view to trapping and trading with Indians.

Not being successful in this enterprise, they turned their attention to prospecting for gold.

They discovered gold on Dease River, a tributary of the Mackenzie, at a place known as McCulloch's bar. Those two men had endured many hardships, such as are only known to the frontier explorer. They had spent three years' time wandering through the mountains, living by hunting and fishing; in all that time never seeing the face of any other white man. When they discovered the gold they were lost, not knowing where they were. Fortunately they met some Cassiar Indians.

Those Indians had visited Stickeen River, where gold had been found some years before and a few white men were working. The Indians piloted them to the white men. The news of this discovery was carried to Fort Wrangel, then a trading post, from thence to Victoria. In those days news of gold discoveries meant a rush, and in the following spring it came. Hundreds went to the new Eldorado. Some were fortunate, but a large majority were not, and returned poorer than they went. The first year in the camp provisions were very high, every eatable one dollar per pound, and very scarce at that.

The summer passed away with us, nothing occurring to disturb the monotony of our lives. We kept on prospecting, but were not fortunate enough to find any paying mines. Our boyish dreams of wealth and pleasure vanished, and we were learning the lessons of life's hard reality. Winter was rapidly approaching. Our purses were light and growing lighter every day. I concluded I had had enough of that country and told my companions so. We bid each other good-bye, and with one companion I started for the coast, which we reached in August.

Chapter II

I HAD HEARD OF A NEW GOLD DISCOVERY on the coast, about one hundred miles north of Fort Wrangel, and wished to try my luck there. My companion and I arrived safely at Wrangel, and at once proceeded to outfit for the new diggings. We bought a canoe, and in company with two other miners from Fort Wrangel, started for the mines. The first day out the weather was very pleasant, and we rowed along, joking and having a jolly time generally. The canoe route was an inland passage. The ocean swells are broken by numerous islands, although in places the passage is twelve or fifteen miles wide.

The second day, in the midst of one of those wide places we were overtaken by a storm. Being unaccustomed to the coast, we were in great danger of being swamped, and could not find a place sufficiently sheltered from the wind and waves to risk a landing. The storm drove us miles off our course, and it began to rain very hard. Darkness came on, and it required the labor of one man continuously at bailing to keep the canoe from filling with water. The darkness was so great, that to see ten feet in any direction was barely possible. We were obliged to judge by the sound of the breakers on shore and keep off, in order to avoid the rocks; for, had we struck a rock on that dark and stormy night, my narrative would end here. Thus we passed the night, wet and cold, for in that latitude September nights are chilly. The gray dawn of morning showed us land about three miles distant, and also what appeared to be a harbor, for which we at once started.

We found the place to be inhabited by Cake [Kake] Indians, a tribe that, although not openly hostile were very treacherous. Several small parties who had been prospecting in their neighborhood had disappeared, and had never been heard from afterwards, although some of their goods such as watches and other articles, had been found in the possession of these Indians.

We had been driven by the wind far out of our course, and to stop there and wait until the storm abated was a necessity. Knowing that they were treacherous murderers, we had before us a very unpleasant prospect. However, we were well armed and on our guard, and, although only four in number, decided to take the chances. They received us in a very friendly manner, and insisted on helping us to unload our goods and pitch our tents.

Having secured our luggage and made a roaring log fire, we cooked our breakfast. Our next move was to get some sleep after our night of exposure. This we proceeded to do, keeping one man on guard, giving as a reason that he was keeping up the fire to dry our clothing. The rain had now ceased. My turn to watch came in the evening, and being very suspicious of them I watched them narrowly. Several of them were loafing around our fire and I observed them eyeing our guns. They appeared to be taking estimates as to the value of our goods, and I was fully satisfied that they intended to attack us in the night. However the storm was abating.

In order that they might not suspect us I quietly consulted with the others, and, our camp being a short distance from theirs, we decided that after they went, or pretended to go to bed, we would slip off under cover of the darkness. This we thought was the most feasible plan to get away, for we were in extreme danger. They numbered at least one hundred, and, although we were well armed, and they were not, we could not expect to stand unharmed against such great odds.

Accordingly we made preparations as if we intended to stay that night, but took good care to let the fire go out so that its light might not betray us, and also so that those of them who were loafing in our camp might go to their own. After darkness had set in, and our unwelcome visitors had all left, we quietly carried our effects aboard our canoe, leaving the tents until the last, knowing that they would watch the white tents, if anything.

All things being on board except the tents, we then prepared to remove them. Some of the party were in favor of leaving them; but to live in that wet, rainy climate without tents was almost impossible, so we concluded to take them with us if possible. One man was left to guard the canoe, two were to quickly take down the tents, and I was to take my station about midway between the Indian camp and my party, and in case of an attack cover the retreat with my Winchester rifle. I crept silently toward the village, until I could see in the darkness the outline of their bark houses. I looked toward my party and the tents and saw their ghostly forms disappear toward the canoe, and had begun to think that we were needlessly alarmed, when an unearthly yell of rage and disappointment broke the stillness of the night. I thought there must be fifty of them at least, and unfortunately for me they were lower down in the brush and almost between me and the canoe.

I felt sure that my companions had reached the canoe with the tents. This was very satisfactory, but my own situation was very unpleasant. However the heathens did not know that I was there, while I knew very well where they were. This was a great advantage to me. But how about my companions in the canoe? They would be obliged to pull away from the shore in order to avoid the villains; in fact I had heard them shove off. The Indians fired several shots after them, which the boys told me afterwards they did not return, as they did not want to expose their position and did not want to go far from shore until they could get me aboard. My mind was quickly made up as to the best course to pursue. The Indians did not attempt to follow them in the dark, although they had plenty of canoes.

As they did not know I was on shore, I was comparatively safe. I was satisfied that the boys would wait until all was quiet, and then return for me. The villains went to where we had camped to see if we had left anything, and then, passing close to where I was hid, went to their own village. Soon all was quiet. I crept silently toward the beach, and it was not long before the canoe came in close to the shore. Some of the boys gave a low whistle, which I answered with due caution. I got aboard, and we were all happy once more. The night was dark, but the sea was calm as a millpond.

We headed for the Shuck inlet,* along which the new diggings were. In the morning we made land side. We landed, cooked our breakfast and felt like new men. Here we remained during the day, resting ourselves from the fatigue and excitement of the

*Actually the Shuck River, which enters the head of Windham Bay on the Alaska mainland.

previous twenty-four hours. The following morning we resumed our journey, which we accomplished without further difficulty.

We found about eight miners only at the diggings, and their reports were by no means flattering, as they said they were hardly getting gold enough to pay them wages. We also found an abominable climate. It was raining when we landed, and the rain fell in that place during my sojourn there, full three-fourths of the time. The cause of this enormous amount of rain in that place is found in the glaciers and ice fields which are found along the stream and inlet, the evaporation from the warm ocean current, which is driven by the wind to the cold icy surface of the glaciers. This causes the vapor to condense, and an almost continual rainfall is the consequence. The warm current which I refer to is called the Japanese current. Drifting up from the South Pacific ocean, along the coast of Japan, it flows eastward along the coast of Alaska, warming the coast and making it habitable, although it makes the rainfall very heavy.*

Nothing daunted by the discouraging reports, we commenced prospecting, and found gold in several places in small quantities. Finally we found a place which we thought would pay. We worked at it about ten days and found it would not pay. The three who were with me then became discouraged and returned to Fort Wrangel, leaving me alone. I then pitched my tent with the other miners who were working on the creek, and continued to prospect on the creek about one mile above my camp.

There I found a narrow canyon about one-half mile in length. Its walls were perpendicular. The miners told me that no one had ever prospected in it, and that it was full of gravel and might contain gold. It could be prospected only at low water. Watching my opportunity I one day found the water low enough to allow me to do some prospecting.

The prospect was flattering. With nothing but my prospecting pan I panned out ten dollars and twenty-five cents worth of gold in two hours. But it commenced raining again, and rain meant high water, so I had to get out of the canyon. Then I turned my attention to some means of turning off the water. I saw that it

*This current actually crosses the Pacific and flows north and west along the Alaska coast.

Placer miners at work on their claims.

would be necessary to build a flume to carry the water over, as the rock was perpendicular on both sides and there was no chance to turn it by.

To build a flume I must have several thousand feet of lumber. This would have to be whipsawed, for there were no saw mills in Alaska in those days. To whipsaw required two men; and I must have help. So I showed the gold I had panned out to two of the old hands on the creek, and gave them equal shares with myself in the discovery. We went to work whipsawing, making boxes and getting ready generally to open the claim in good shape. It took us three months to make the necessary lumber sluices, and another month to put in the flume.

Our provisions ran short, but game was shorter. We killed a pair of black bears, but they were scarcely fit to eat. They tasted very strongly of fish, living, as they did at that time of year, on salmon that were ascending the river to spawn. However fish were plentiful, and I sent a white man and two Indians to Fort Wrangel for supplies. Our daily fare was fish and bear for breakfast, bear and fish for dinner and both together for supper.

Having completed our flume and got everything in readiness, we commenced washing only to meet with disappointment. The claim did not prove to be near so rich as the prospect promised, and would barely pay for working, leaving no surplus to pay for the months we spent in preparation. However we worked away all winter, being obliged to stop occasionally on account of spells of cold weather.

The man I had sent for supplies arrived, but had got caught in a storm and had to throw half the cargo overboard in order to keep from swamping, so we were short of a great many things all winter. We killed some mountain sheep which helped us out materially.*

The mines kept yielding less and less, until they would no longer pay running expenses. Spring came, and with it high water, which washed out our flume and we abandoned the claim. We returned to Fort Wrangel where each took his own course.

---

*Probably mountain goats.

Chapter III

I CONCLUDED TO SPEND THE SUMMER of 1878 prospecting on Prince of Wales Island. Copper had been discovered on this island, and the formation was also favorable for the more precious metals. The island was inhabited by the Hyder [Haida] Indians, a tribe noted for their intelligence and good-will toward the whites. Accordingly, in company with one companion, I started for Carter [Karta] Bay, distant from Fort Wrangel about ninety miles. We arrived safely, and were welcomed by the natives in a very friendly and hospitable manner. The chief had a house made of lumber, which he had split out of cedar trees. It was roofed with bark. He tendered us a corner of his house to live in; and as it was clean and exceptionally well kept for an Indian's house we accepted it, making him presents in return, of tobacco, coffee, and sugar.

These people are great canoe builders; the timber, which is red cedar, growing quite large on this island. Some of their canoes are models of beauty, and some of them are very large. I have seen them made from the trunk of a single tree so large that they could easily carry eight tons.

These natives, when not engaged in hunting or fishing, spend their time in making those canoes, both large and small, which they trade or sell to neighboring tribes, who have not the time or the skill to make them. They told us that they would be pleased to help us to find gold or other precious metals on the island, for they wanted plenty of white men to come and live with them so that they could trade with them for flour and other articles.

They materially assisted us in prospecting, carrying us from place to place in their canoes, and guiding us inland to places where they thought we might be successful. Also, they brought us fish and game, for which their charges were very moderate. But I could not find even a trace of gold. In one place near a good harbor I found several copper leads, and the ore looked promising. As it might carry silver, my partner and I did considerable work here. After seeing that the leads had a permanent appearance, the next thing was to find the value of the ore.

This could only be accomplished by taking some of the ore and going to San Francisco, and having it properly tested. We decided that my partner should go. So packing up a quantity of the ore, I took him and the ore to Pongas [Tongass] Narrows, the nearest point where he could board the mail steamer. Having seen him off to San Francisco, I returned to Prince of Wales Island, and continued to do what I could to develop our property.

In the meantime I was alone among the Indians, there not being a white man nearer than Fort Wrangel, ninety miles distant. Two months elapsed and no partner, and no letter, for I had no chance to send to Wrangel the nearest post office. Another month and no news. My provisions were almost gone, so I hired a small canoe and one Indian and started for Fort Wrangel for supplies.

Here I met with a little adventure that almost cost me my life. One day while pulling along in our little canoe close to the shore I saw some berries on the bank, a species of the whortleberry, which grows in great quantities on the island. I told the Indian, who was steering, that we would go ashore and get some of them. We landed and at once commenced eating berries. The tide was rising and we had not properly secured our canoe, which got afloat, and when we went back to the place it was several hundred feet from land. Here was a fix. I could see only one way to get it — to swim for it. Removing my clothing I plunged in and struck out boldly for the canoe. I had got within about fifty feet of it when a wind sprung up from off the shore and blew it away faster than I could swim. In vain. I did my best, but I could not gain one inch. I looked back towards the shore. I knew I never could swim back, as I was almost exhausted. It was a bad position to be placed in. I determined to make one more effort for the canoe, and at least not to give up as long as I could keep above the surface. I kept heading for the canoe. I was so exhausted that I thought every stroke would be the last I could make, when I thought the wind slacked. Then I noticed that I was gaining on the canoe. This encouraged me and I kept on. The wind slacked entirely and I gained the canoe, crawled in and lay in the bottom quite a while before I got strength enough to pull ashore.

Indian dugout canoes.

I have often thought to what apparently trifling circumstances we owe our lives. For had the wind continued to blow a few short minutes longer, my doom would have been sealed. I could not have gained the canoe, and I should not have had the strength to return to land. But my time had not come. The All Powerful had interposed, and I lived on and have passed through many hardships and dangers since. Although I am yet a young man, and should only be in the prime of life, I am physically a wreck through cold, starvation and exposure, in the land of the midnight sun. A few days afterward I reached Fort Wrangel.

Here I received mail from my partner. He had had the ore tested and it proved to be worthless. It carried some copper, but not enough to make it valuable; no gold or silver. He did not intend to return until the coming spring. This discouraged me from prospecting further in that locality. I went to a bay on the west side of the same island and spent two months, but found nothing. The balance of the time, until February, 1879, I spent hunting.

Chapter IV

RETURNING TO FORT WRANGEL I prepared to sled up the Stickeen River on the ice, for late in the fall a new gold-bearing creek had been discovered in the Cassiar district, and I wanted to try my luck on it. All things being ready I started with one companion up the river. We were taking considerable supplies with us, which, in that country in winter or early spring, are always conveyed on hand-sleds; some times with dogs, but oftener the miners not having dogs pull them themselves. My partner and I not having dogs were doing our own pulling. We had too much freight to take it all at one load, and were obliged to make two trips.

Thus, suppose we had a thousand pounds, we would take five hundred and haul it one half day's journey, and go back in the afternoon. The next day we would haul up the rear load, and so on until the journey was finished. A good many parties were ahead of us. Thus the road was broken and very good, though the weather was cold. All went smoothly enough, until one stormy morning we had a longer journey than usual to make. The snow was drifting some, which delayed us, as it made the sleds harder to pull, and it was almost dark before we got started back to camp where our beds were.

The storm increased, and to add to our discomfort it was bitter cold and blowing in our faces. It became very dark, and in many places the trail was drifted full, as the wind having blown the snow all clean from the glare river ice left us no guide to go by.

In all those northern rivers that are frozen in winter, there are air holes or places that do not freeze, caused by currents of air. I had thought of those places, but had not seen any as we came up during daylight; but we had lost the trail and in the darkness got off our course. All of a sudden my footing gave way and I was floundering in the cold river water. My partner, who was behind, warned by the splash I made managed to stop, but not before he got partly in. I had fallen headlong. Fortunately the water was shallow, not over three feet deep. I regained my feet before I got drifted under the ice, and climbed out again on the solid ice. In a moment my clothes had frozen stiff; and to add to the horror of the situation, we were lost in the darkness and did not know where to go.

My partner had only got his feet and legs wet, as he had not fallen down. To attempt to find the camp in that bitter cold wind, I knew was almost sure death. So we headed for a bunch of timber on the bank and penetrated it until we found a sheltered place. We tramped a path in the snow around a large cottonwood tree, and kept up a kind of Indian war dance around it. My frozen pants broke through at the knees, and I saw at once that we must do something else, for it was early in the spring, and we could not stand it all night, for the nights up north in February are long. As I had got wet all over, I had no matches that would burn. Fortunately my partner had some that had not got wet. We went in search of, and at length found a dry stub of a tree, and after a great deal of trouble and when our matches were almost gone, we finally got it to burn.

It made a good fire and burned all night and saved us from freezing; if not our whole bodies, at least some of our limbs.

At daylight we found we were but a short distance from camp. We went to camp, made a roaring fire, got a good breakfast and slept all day.

I have dreamed of that plunge in that cold water on that cold night months afterward, and woke up with a warning cry to my partner, that seemed to me almost as real as it did on that wretched night. But such is the life of the miners and frontiersmen of the far north. They are exposed to cold, hunger, wild beasts, savage tribes and all diseases induced by their rough mode of living, many even dying with scurvy.

All this is borne without complaint; and a miner will divide his last morsel with a comrade in distress. To abandon a sick or wounded miner is looked upon as the most unpardonable of crimes. You must stay with and nurse him, or bring him to a place where medical assistance can be procured. I know men who have assisted in carrying or hauling hundreds of miles.

Having recovered from our night of misfortune, we proceeded on our journey. We were more fortunate the balance of the trip. Nothing occurred to delay or annoy us. We arrived safely at the diggings and found the weather cold and everything frozen solid. We went to the new creek, staked off claims, made ourselves comfortable, and waited for the weather to moderate so that we could prospect.

In a few days the weather moderated and we went to work prospecting. We sunk holes to the bed rock in various places. This was quite a job, for the ground was frozen solid all the way. We found some gold but the mine was not rich. We concluded to work it, and prepared accordingly. We sawed lumber, made sluice boxes, and built a wing and head dam. Having got the water turned off, we commenced washing. The claim paid well, and we flattered ourselves that we should do well that summer. It paid for about one month, and then all at once it failed. Although we spent time and money trying to find the pay-streak again we did not succeed.

We sold out to some Chinamen, for the claim would still pay enough to satisfy a Chinaman. We packed our effects and started for new fields. But I did not find anything more that summer although I was busy prospecting. One day I was prospecting in a small gulch. I sunk a little shallow hole to the bed rock, and filled my pan from the bottom of the hole. I went to the water and washed it out in that pan. I got a nugget that was worth twenty-eight dollars, but strange to say, I could not find another in the gulch. I often wondered how that nugget got there all by itself. I continued to prospect until the snow came, but met with no reward. Then I returned to Fort Wrangel in the fall of 1879.

Fort Wrangel was one of five posts established in Alaska by the U.S. Army soon after the transfer of the Territory from Russia. Built in 1868-69, Fort Wrangel consisted of a number of log buildings surrounded by a stockade and it stood on the site of the present federal building at Wrangell. The Army occupied the fort off and on from the time of its construction until the troops were withdrawn from Alaska in the summer of 1877. The stockade soon disappeared but some of the buildings survived for many years and were used for a variety of purposes.

Chapter V

WHEN I ARRIVED AT FORT WRANGEL, I found that there were very favorable reports from the Sitka quartz mines. Several years before this, a practical quartz miner had enlisted in the United States army, and with his company had been assigned to duty at Sitka, Barnoff [Baranof] Island, Alaska. Now being something of an expert in quartz, he soon discovered that the country was gold-bearing. At the expiration of his term of service he remained in Alaska to prospect. He succeeded in making several valuable discoveries. He went to Portland, Oregon, organized a company and commenced work on the mines. This is what caused the favorable reports I have mentioned.

I took the first mail steamer that came along for Sitka.

The site of the present town of Sitka is not where the Russians first settled. They first settled in a place called at present Old Sitka, where they lived and traded with the natives a number of years. But finally the natives became hostile, and one dark and stormy night treacherously gained entrance to the Russian fort and murdered every man, woman, and child, leaving no one to tell the tale.

When this became known to the Russian government they promptly sent a war ship, the natives were severely punished and a new town was started on the present site of the town of Sitka. I am not able to state how long since this occurred, but it was a great many years ago, for the new town of Sitka has a very ancient appearance. Here the Russians built very large and substantial buildings of hewed logs, or square timber. At present some of them are rotted to the ground, and all are more or less rotten and dilapidated. Here I found a couple of hundred Russians, or people of Russian descent; also many half-breeds. They were of the lower class of Russians, and did not appear in the least superior to the native Indians. They were drunken and debauched in the extreme. It was a common sight to see a Russian

Looking east on Lincoln Street in Sitka in the 1870's, with Saint Michael's Cathedral in the center — although the engraver neglected to put a cross atop the spire. The cathedral was destroyed by fire on January 2, 1966, but has been rebuilt. At the extreme right in the picture is a log building with a sign, A.T. Whitford, over the door. Whitford became a partner of John G. Brady in the Sitka Trading Company and the building, torn down around 1920, was long known as The Old Trading Post. Only one of the buildings shown here, the one with the chimney, is still standing and it is one of the last Russian log buildings in Sitka.

woman going along the streets staggering under the influence of whiskey or some other vile compound, using the most profane and indecent language and cursing some one for some real or imaginary wrong.

Their living is principally fish, or the same as that of the natives. They seldom have flour or groceries, for all their money goes for alcoholic stimulants, and it is fortunate for them that they have no opportunity to get much money. Here both Russians and Indians manufacture by distilling a liquor called hoochenoo. It is made from molasses, sugar, potatoes and in fact from almost anything that will ferment. They make their stills from old tin coal oil cans, also making the worm out of scraps of tin soldered together. At this business they display considerable ingenuity, and it is a noticeable fact that a native Alaskan, whether he be of Russian or Indian parentage, will bring forth all his mechanical skill if there is any whiskey to be got by it. Such is the life led by these poor, miserable people. Yet nature has furnished them with food in abundance. The waters are teeming with fish, which are easily taken at all seasons. The land supplies them with deer, bear and other smaller game, so that with the necessaries of life they are well provided for; much better than thousands of the poor of Europe and the larger cities of America, where, in cold winter, children roam the streets, selling papers, matches and other light articles to get the money to buy their daily bread.

No native child in Alaska is so poor. Their houses are stocked with dried fish and game. Fuel is abundant, and they are happy. If they would continue to live according to their ancient customs they would remain happy, but they are trying to imitate the whites. In this they are encouraged by the missionaries and the United States government.

It is a mistake, and this is the reason why: The government has appropriated certain sums yearly for the education of the native Alaskans. The money is handled by a mission, and a sectarian mission at that. This mission also adds to the amount by private subscriptions through the States, and the two sources of supply combined make a considerable sum. With this money they have built a large boarding school at Sitka, and also other schools at less important places.

At Sitka they gather in a large number of Indian children. More girls than boys enter the schools; but some of the boys not

Russian Orthodox church at Sitka.

liking the restraints run away. Thus the schools are composed mainly of girls. The boys and girls that remain are fed and clothed on the civilized plan. They acquire tastes for the white people's style of living. They are taught to cook and sew, and keep themselves clean. They are also taught to speak English, to read and write; in fact all the branches of a common school education. They are also taught the knowledge of the Scriptures. Now this is all right; that is, the education and the religious teaching. But not so the boarding house arrangement. There the females acquire the same tastes that our own mothers and sisters have. They live on the same food, cooked in the same style that the whites use. When their time expires at the school they are sent out to make their own living. They have no knowledge of the ways and means pursued by their fathers and mothers in getting their support from the sea and forest. They cannot get employment from the whites, and they are truly in a helpless condition. You would naturally ask the question: Why do they not marry the young men of their tribe? The reason is this: The men are raised as Indians, the females with the habits of the whites, and they have no tastes in common. Their boarding school education estranges them from their natural protectors. This state of affairs is taken advantage of by white men, men of bad principles. The girls are ruined body and soul. A few short years and intemperance and other vices have done their deadly work, and the chapter of their life is closed forever.

Such being the case, why would it not be better to drop the boarding school system. To educate them is all right. To teach them the laws of God is also good. But it is self-evident that to teach them a mode of living, which the natural resources of the country will not support, is a mistake. It is also a noticeable fact, that the most criminal and vagabond class among the natives are those who have been educated by the whites. This is a deplorable state of affairs, yet the officers of the government do not appear to take any notice of it. And it is not likely that they will so long as their own fat salaries are promptly paid.

The natives are rapidly decreasing in numbers. This decrease is caused by the ravages of diseases that are always the result of intemperance and other vices. The death rate among them at present far exceeds the birth rate. A few short years more of the present style of management and the native Alaskan Indian

Three Russian-built log buildings in Sitka in the 1870's, when the U.S. Army was still stationed there. The upper building was known as Baranof's Castle and was built in 1837 and destroyed by fire in 1894. To the right of it is the tall flagpole upon which the Stars and Stripes were raised on October 18, 1867, showing the guylines upon which the Russian flag fouled as it was being lowered. Below the castle is a building used by the Russian-American Company as offices, with a school on the second floor. It became the U.S. Custom House and was occupied by other federal offices as well, including the post office. It was destroyed by fire early in 1936. The three-story building at the right was used as a barracks by the Russians and then, under the American flag, by first the Army, then the Marine Corps. It also housed the U.S. District Court of Alaska for some years, and the federal jail. It was torn down about 1920.

will be numbered with the races that have passed from earth never to return.

The Russian population of Sitka, with a few exceptions, are not at all superior to the Indians, except in their origin. Their habits of life are the same. The morals of the Russian Alaskan are no better than those of his darker skinned neighbor, and it was a common sight to see them band together in drunken debauches. They also are decreasing in numbers. Some of them will work when they can get employment. So also will the Indians, who do almost all the packing and carrying of supplies for miners and others through the territory.

Chapter VI

BUT NOW, DEAR READER, having given an insight into the habits and morals of the people of Sitka, I will proceed to speak of its industries. The mines are situated twelve miles from Sitka, near the head of an arm of the sea, called Silver Bay. A small steam launch was used by the company to transport the mining supplies to the head of salt water. From thence to the mine, about two miles, Indian labor was used. The quartz was gold-bearing, carrying no other minerals of any value. It was also of the kind that some call ribbon rock. Portions of it were very rich in gold, while the greater part was almost entirely worthless.

Our boarding house was at an elevation of twelve hundred feet above sea level, and in that high latitude twelve hundred feet high means cold weather in winter. Our quarters were made of lumber, with the cracks battened up. It was really very poor quarters for so cold a climate in winter. Often the victuals would freeze on the table while we were having our meals. The company were very anxious to get their little five stamp mill built in order to test the value of their mine, and would not stop for cold weather. Eventually, by perseverance, we got the machinery hauled over the snow on sleds, employing Indians to haul the sleds. We built the mill and drove a tunnel, and by the first of March the following spring, 1880, were ready to run. The company had expended eight thousand dollars. All things being ready we started up, only to find that the ore was not rich enough to pay. The work shut down. The money expended was a dead loss, and the mine is there for any one that wants it.*

*This was the Stewart Mine which a German-born mining engineer, George Pilz, was attempting to open for the owners. It was the first stamp mill in Alaska.

Chapter VII

THE SUMMER OF 1880 I SPENT IN PROSPECTING in the vicinity of Sitka. I did not find anything of any value. I was thinking I would leave Alaska, and was waiting in Sitka for the ship. We then had only one vessel a month. Just then two other prospectors came in, who had been prospecting in the neighborhood of Tarko [Taku] River. They had made good discoveries, both in quartz and placer gold. I concluded to remain, and joined a party with four others.

We purchased a boat, and taking about two weeks' provisions with us started for the new discovery. It was our intention to go there as soon as possible — try to get claims near the discovery and return to Sitka to winter. It was then the first of November, and we did not care to try to do anything during the winter.

We were not acquainted with the coast in that locality, and were obliged to travel by an old Russian chart, which I had managed to pick up at Sitka. The chart was not correct and caused us to lose our way several times, for our route lay among the numerous islands along the coast. After a good deal of traveling and hunting around we succeeded in finding the Auk Indian village. The Auks are a small tribe of Indians who live in the vicinity of the place where the gold was found.

I have neglected to mention that we could not get any reliable information from the two men who made the discovery, as they did not want parties to go there before spring, fearing that they might interfere with their claims. Had they been willing to give us a map of the location it would have saved us much time and annoyance. However the Auk Indians knew the place, but were smart enough to demand a large fee for showing us where it was. It also began to snow, which made it yet more difficult, as it covered up the tracks made by the first parties.

The Indians demanded one hundred dollars for showing us the place. This demand we did not feel inclined to comply with. After a good deal of dickering and losing several days more time, we finally induced them, for a small sum, to show us where the discoverers had camped on the beach. Every thing was covered with snow, but we kept up our search and finally found their location stakes in a little valley about four miles from our camp.

Owing to the depth of the snow, which was increasing every day, it was impossible to prospect to any advantage, and we were forced to give up trying to get any claims until spring. In the meantime our provisions had completely given out. We were one hundred and sixty miles from Sitka, the nearest supply point.

We decided to return at once to Sitka. First we went to the Indian village where we invested several dollars in dried salmon, which, although not very palatable food, is a great deal better than nothing.

We started for Sitka, but the winter storms were fully upon us. The wind blew continuously, and, day after day, we were obliged to lie in camp, the wind and seas making it too dangerous to travel. Our food was dry salmon and clams; and to add to our discomfort the snow turned to rain, and the heavy winds drove it through our tent like a sieve. All of our clothing and blankets were very wet — so wet that we could not sleep in them. We could not dry them for the rain was continuous and heavy, however, we managed to make a large log-heap fire. Fortunately we found some pitch pine which made a fire possible.

This added much to our comfort. Here we could cook three meals a day, and have a change every meal. Thus, salmon and clams for breakfast, clams and salmon for dinner, and reverse it again for supper. On several occasions I went hunting, and once I was rewarded with a porcupine. He was fat and made us a fine stew.

Well it could not storm always, so one day it quit and cleared off. We launched our boat and returned to Sitka without further trouble.

By this time a lot of miners, who had come from the Cassiar mines to the coast to winter, heard of the new strike and had come to Sitka to outfit and go there. This created a stampede. Every one wanted to go. Although nothing could be done during the winter, they would be there when spring opened. So everybody, or all the miners, went there.

Again I outfitted. This time with two partners, and taking winter supplies went back. The weather was good, and we had no trouble on the trip. We all camped on the beach, about forty miners all told, and started a new town. That town is now Juneau City, the metropolis of Alaska.

Chapter VIII

WE SPENT THE WINTER in building log houses and sawing lumber with whipsaws, for we had a vague idea that we would want the lumber in the spring. Miners came from other places, and when spring opened there were nearly two hundred miners on hand.

As the snow melted away all were busy prospecting, and almost every one secured claims of some value, both of placier [placer] and quartz. Some of the quartz was very rich; not so with the placier, although it paid to work it, that was about all. It did not prove to be rich.

Early in the spring the Douglas Island mines were discovered, among which is the famous Treadwell mine. This mine was discovered by an Indian, who showed it to a Frenchman. The Frenchman located and sold the mine to John Treadwell & Co., for four hundred dollars. Today it is the greatest mine in the world.

The company reside in San Franciso. They have a two hundred and forty stamp mill, with a sulphuret concentrator to each battery of five stamps. They also have large works for extracting the gold from the sulphuret. The mine is the largest known body of quartz in the world, the ledge being about five hundred feet wide, cropping out this width for thousands of feet in length. There are other large and high grade ledges of high grade ore, which will undoubtedly become paying mines when properly opened and worked.

The Juneau City mines, and Douglas Island mines are in the same district, about five miles apart. The Douglas Island mines are situated on Douglas Island, directly opposite Juneau, across a narrow channel of salt water and about two miles distant from Juneau City, while the mines on the Juneau City side are located all the way from the town to a distance of four miles back from salt water.

Many of the quartz mines here have great natural advantages for shipping machinery. It could be put aboard the ship in San Francisco and landed right on the mill site in Alaska, where it would be required for use. Such is the situation of the famous Treadwell mine.

Labor here is not high, considering the climate and other disagreeable surroundings, which the laborer has to contend with. Wages range all the way from two to six dollars a day. A common white laborer will get three dollars per day, a miner four, a blacksmith five, a millwright six and an Indian two dollars per day.

The Indians are largely employed in packing. This they generally do by contract, and are paid according to the distance and the weight of the article to be packed. They are very treacherous and untrustworthy. I have known them to agree to pack a certain distance at a certain price; they would start with the load and after packing it a short distance, and yet so far from town that you could not get any one else to pack it, because there was no one there to get, they would set the load down and demand double the price agreed upon. If you would not pay it they would leave you and your goods on the road, leaving you in an unpleasant situation. You could not carry the load yourself, and if you left it, it would be stolen, and there was no one near whom you could get to pack it for you. This was very annoying. Yet to all these inconveniences we must submit. All for the sake of the poor Indian.

I have also seen the members of one tribe prevent the members of another from packing or working for white men, because they did not think the pay sufficient. There the Indians come in great numbers in search of employment. Being of different tribes, they do a great deal of quarreling among themselves, and even go so far sometimes as to kill each other.

However, when it got so far as that, it became necessary for the whites to interfere, for we wanted the Indians to labor, and if they were fighting among themselves we could not get them to work.

Unfortunately among the white population there were some unprincipled scoundrels, who made it a business to sell vile whiskey to them, and as we had no law it could only be prevented by the miners making a rule forbidding it, and as we had made our own mining laws we also added a clause to that effect. Yet there were some who paid no attention to this, and on several occasions it became necessary for the miners to seize and destroy their poisonous mixtures.

One of this class of men located himself on a trail leading from Juneau back to the portion of the district called the basin. This trail was largely traveled by Indians, and the reprobate had built a log cabin, where he was doing a rushing business. One evening three Indians came along who were intoxicated before they came to his place. They had no money, but demanded liquor. He refused them credit — a fight was the result, and the white man was killed. One of the natives struck him on the head with a club and broke his skull. The Indian who did the deed was one of the most lawless of his tribe, and there were strong suspicions that he had killed white men before. The next day after committing the murder, he asserted that he would kill more white men yet.

Now the Indians had been very saucy and impudent all summer. There being a great many more Indians than white men in the camp, they had received the impression that the whites were afraid of them. So a citizens' meeting was called, and it was decided that it would not do to let the murder go unnoticed. It was thought best to arrest him and turn him over to the United States authorities at Sitka.

As we knew that the gunboat would be over from Sitka in a few days, the arrest was made at once. He resisted and was aided by the two who were with him when the murder was committed. This made it necessary to arrest the three. A man was appointed to guard them and they were locked up in a building that had two rooms. The guard occupied the front room and the prisoners were locked up in the back room. They were confined here about one week, when one of the prisoners pretending that he wanted to do some sewing, asked and received permission from the guard to sit in the front room, as the back room was too dark, there being no window. The guard was armed, but foolishly and carelessly had allowed his pistol to lie on the table in the room. This was what the murderous, treacherous Indian wanted, and watching when the guard turned away for an instant, he seized the pistol and shot him in the back. Then unbarring the door released the other two, and all three ran off to the Indian town, taking the pistol with them.

The Indian town is situated about one-half mile from Juneau, and at the time when this occurred there were at least one thousand Indians in the village. The murderous act was partly witnessed by a miner who was near the building. He foolishly, alone and unarmed, followed the three fugitives who ran into an Indian house. He attempted to enter also, when the Indian who had the pistol shot him dead at the door.

At once all was excitement, both in the Indian and white settlements, and a conflict appeared to be inevitable. The miners were notified in every direction and all came to town. Fortunately we had plenty of arms and ammunition. The Indians far exceeded us in number, for we could not raise more than two hundred and fifty men, while the Indians could raise a thousand. Yet we had the advantage of being well armed, while they had nothing but old style shot guns. Still it was our desire to avoid a conflict, as we could not profitably fight Indians and mine at the same time. For in case of a continued war much property would be exposed to their depredations, and also many isolated parties of miners who were out prospecting, might be killed. Yet it was necessary to punish the murderers, as it would not do to let them commit such murders and go free. A small detachment of armed men was sent to the Indians to demand the prisoners. We found the Indians all armed, but they did not appear to wish to fight. We called for their chiefs, who came forth, and we spoke to them, telling them that we must have the prisoners; that we did not wish to fight with and kill others of them who were innocent of crime; that we only wanted those who had murdered our friends; that we wished to remain on friendly terms with them; that if they did not give them up quietly we should be obliged to take them by force, and that we would be assisted by the government, who would undoubtedly send gunboats to help us. This had the desired effect. If you want to scare an Alaska Indian talk gunboat and cannon to him. After considerable talk amongst themselves they agreed to give them up. The chief pointed out the house they were in and told us to go and get them, that they would not interfere.

We found the door barred, and they would not answer when we asked them to open. Procuring a large stick we quickly battered down the door and entered. They had got a board loose on the back side of the hut, and, jumping through, attempted to escape. But escape was impossible, as the house was surrounded.

The Indian who committed the first murder, still having the murdered guard's pistol with one load in it, placed the muzzle to his temple and blew out his brains, thus cheating the gallows. We secured the other two and took them back to Juneau, passing through almost the entire length of the Indian village. On our return we passed by groups of natives. They had put away their arms, but looked sulky and threatening.

A meeting of all the citizens was at once called, and consultations had as to what should be done with the prisoners. It was decided to at once appoint a judge, impanel a jury and give them as fair a trial as possible in the absence of proper authority and law. This was done. Witnesses were examined, the evidence was positive and very strong. The verdict of the jury was: "Guilty of murder in the first degree." The sentence was death.

A committee was at once appointed to build a scaffold. It was built on the beach, in plain view of the Indian village, in order that it might be a lesson to them. All things being completed, the prisoners were brought forth under a strong guard. It may be said to their credit that they showed no signs of fear. They both made short speeches in which they expressed themselves willing to die, and they had confidence that they were going to a happier land. They advised their friends to live at peace with the white man and conform to his laws; also to avoid the use of whiskey and other intoxicants. They said that they forgave their enemies and wished all to forgive them. Having spoken thus they pronounced themselves ready. The black cap was put on, the noose adjusted, the trap sprung, and there were two more good Indians on earth and two more hunters in the happy hunting ground. When they had hung long enough, their bodies were cut down and taken away by their relatives.

This ended the unfortunate affair, but it had an excellent effect on the natives afterwards. They were quite civil to the whites, and although this occurred several years ago, there has never been any trouble with them since.

Chapter IX

IN TIMES BEFORE THE ADVENT OF THE WHITES in Alaska, it was the Indian custom to hunt and fish in summer, laying up great stores of dried fish and game for winter use. Their trapping for furs was done in winter. The Russian Fur Company had established trading posts all along the coast, from the mouth of Stickeen River in the south, to the mouth of the Yukon River in Behrings Straits [Bering Strait] on the north. There was no opposition, and the natives were compelled to take just what the company chose to give them for their furs.

It is needless to say that the prices received by the natives were very low. For example, they would trade an old fashioned flint-lock shot gun for furs after this manner. They would hold the gun upright on the floor, and the Indian who bought it had to pay for it a pile of marten skins reaching up to the top of the gun. Every skin would be worth five or six dollars in the home market. Thus they realized five or six hundred dollars for an article not worth ten. All their other furs they were cheated out of in the same way.

Yet they were really better off in those days than they are now, for now they have almost quit fishing and hunting. They depend on getting their winter supplies by working for wages for white men in summer. Although they accumulate considerable money in summer, they do not take care of it, and the early winter finds the majority of them destitute. Many of them suffer with hunger, even in this land of abundance, through their own negligence.

Whiskey has much to do with this, for there are still those here, who will furnish them with it, and an Indian will brave hunger and cold at any time if he can only get whiskey.

There is also another custom which is detrimental to their general welfare. If an Indian accumulates more than he requires for himself and family, which is often the case, he gives what in our language would be called a grand ball. He spends all his surplus wealth for groceries and dry goods, and speaking in his language, gives a potlatch, that is, divides the goods up in presents among his guests, often tearing blankets in small strips in order that the division may be equal.

These potlatch parties always leave the giver destitute, for he invariably gives away all he has. In return for this he receives the title of chief, and the greater the amount he has given away the greater chief he is supposed to be. Yet it is only a title, for I never could see that a chief had any authority worth mentioning. All these things keep the natives poor. In times gone by he was happy. He hunted and fished for his food, and trapped for his clothing. He had never heard of an intoxicant of any kind. His health was good; and all things which his natural taste craved, were easily procured. But alas for poor Lo! Those happy days are past and gone. He no longer fishes and hunts. He wields the axe, the hammer, the pick and shovel, and packs huge burdens on his back. He is a slave to the white man and his own depraved appetite for strong drink.

The general health of this once happy people is destroyed; their morals have not improved by contact with the whites; the poison of unmentionable diseases has impregnated the whole race. A few short years more at the present death rate among them, and the copper colored native Alaskan will only exist in history.

The question naturally arises, why is it so? It always has been so where the two races come in contact. One cause is the bad class of white men who are always flocking to the frontier. I do not mean to assert that all are bad; that half, or even one-fourth, are bad. But there are always enough to do irreparable damage by teaching them all the vices practiced by the whites. The government is slow to act; and when it does act, takes little or no interest in the welfare of the natives. Some will ask, what are our missions doing? They are also slow to act, and generally follow up the prospectors and miners, and the seeds of vice are sown before they get there. They should be the first on the ground, because, according to the proverb, an ounce of preventative is worth a pound of cure. But they are too late. The mischief is done, and the confidence of the natives in the white man's honesty and truth is destroyed before they arrive.

A Tlingit Indian dance.

No doubt the intentions of the missionaries are good, but their efforts are sometimes worse than useless. It is so in the case of the young Indians' boarding school, or so called home, at Sitka. The intentions of the supporters of that institution are good, but they are eastern people and do not understand the Indian character. Schools are good, churches are good, but the natives have homes better suited to their character and tastes than can be furnished by the whites.

Chapter X

CONSIDERABLE PROSPECTING is done every summer along the coast, and some valuable discoveries have been made, both of gold and silver bearing rock; gold quartz also exists in large quantities on Admiralty Island.

The prospector in the early spring finds it necessary to exercise a good deal of caution on account of snow slides. On one occasion my companion and I were traveling along the mountain side. We both had packs on our backs, for it is necessary there to carry both your bed and provisions. When traveling in the mountains, if you cannot get an Indian to pack for you, you must do it yourself; for pack animals are out of the question. The hills are too rugged, and the timber too thick. We were at an elevation something like twelve or fifteen hundred feet above the sea, and close to the beach. We came to a field of snow which lay in a declivity on the mountain side. It was about one-fourth of a mile in width and sloping down quite steep to a precipice, or sheer descent of a thousand feet or more. We were obliged to cross this snow field or climb much higher to get around it. Being tired we did not wish to climb, and decided to cross it where we were. We did not like it, but overcoming our fears we started.

In the middle of the snow was a large rock, which was solid in place and bare of snow. My companion was ahead. We had not gone far when I thought I noticed the snow moving. I called to my companion to come back. He laughed at me and kept on. I stopped, when all at once, with a dull, grinding sound, the whole field started over the cliff. To go over that cliff on the snow would be to fall more than one thousand feet into the briny ocean. Drawing my sheath knife I hastily cut the fastenings of my pack on my back and ran for life to the side I started from, leaving my pack to go over the cliff. I then turned to look for my companion. He had fallen down and was struggling to regain his feet. His pack was still on his back. I called to him to cut it loose, but the roaring of the rushing and grinding snow prevented his hearing me; and although he regained his footing he was being rapidly carried to the precipice. Fortunately the sliding snow was carrying him toward the large rock in the middle, which was solid and higher than the surrounding snow. He saw the rock and barely managed to gain it. He remained there till the snow had all passed by. Such is the life of the gold hunter in this far northwest country. Had it not been for the rock in the middle his death would have been certain; for there was no way that I could render him the least assistance. Having lost my pack, which contained some of the necessaries of life, we were obliged to return to our camp on the beach.

Chapter XI

HERE WE REFITTED AND STARTED AGAIN; this time for the interior of the island. We found it very hard traveling, for it was raining almost continuously and the growth of underbrush was so dense as to be in places almost impenetrable. Here grows a shrub called Devil's Club. It grows very dense and is covered with thorns. These thorns are strong and sharp, and long enough to penetrate your clothing. They are slightly poisonous and cause inflammation. In some places are great tracts of fallen timber, and the streams were high and difficult and dangerous to ford.

It is only where the streams of water cut through that the geological formation can be seen. The growths of timber are principally of the coniferous order, interspersed in the low lands with alder, soft maple, and a little cottonwood. Although there is plenty of timber, the quality is inferior. It is scrubby, knotty and wind shaken. It can be used for the rougher building material, but finishing lumber must be shipped from Puget Sound.

The wealth of Alaska consists of its gold mines and fisheries; and the time is not far distant when it will take the lead in those two sources of profit. It will be a profitable field for the prospector for a hundred years to come, and equally so for the fisherman.

The same description in regard to climate and vegetation will apply to both islands and main land in Alaska.

After making several trips like the one described, and boating all around the island, finding nothing of any value, we left it in disgust and tried our fortunes in Berner's [Berners] Bay on the main land, about fifty miles north of Juneau.

Chapter XII

IN THAT LOCALITY GOLD HAD BEEN FOUND, and indications were good for paying strikes. There we made some discoveries, and also some very good prospects were found by other parties; and now there is a promising, but undeveloped mining camp at that place.

Upon one occasion, while prospecting in this district, I had the good fortune to witness a battle between two monstrous male cinnamon bears. Although the fighting of animals has always been something that I did not like to witness, yet I can say it was not the case that time. I do not like bears, especially the large, savage kind; and while they were fighting I was hoping that they would succeed in killing each other. My companion and I were climbing up the mountain side in search of quartz ledges, when our attention was drawn to a very musical noise. It consisted of a continuous growling, grunting and half barking noise, very loud and distinct. At first we did not know what to think of it, but finally concluded it must be bears fighting. So we changed our course in the direction of the noise. We had not traveled far till we came to an open glade, or space of land comparatively level, on which no timber grew — a space of three or four acres — and here were the bears. We did not enter the glade, as we did not want any hand in the fight.

Our first move was to find and climb a good tree and get a secure position thereon, where we could have a good view of the combatants. It was a terrible fight. They bit each other and struck terrible blows and at times appeared to almost take the position of pugilists. While looking on I thought that a human pugilist would stand about as good a chance with them as a mouse with a mastiff. Sometimes they would be striking and warding off terrible blows making fur and blood fly. Then they would close and wrestle and bite and roll on the ground roaring and growling continuously. They were about equally matched, and for a long time neither one appeared to gain any advantage. I never saw anything that looked near so savage as they did in their anger. Finally one of them appeared to be giving out, and having closed again he was thrown and lay still. Taking careful aim with my Winchester rifle at the victor, I fired. Bruin fell beside his vanquished companion, but quickly got up again and started for the timber. As he did not see us in the tree his head was now turned from me; and although I gave him two more shots, he ran away. His opponent meantime lying still did not appear to notice the shots, my companion taking my rifle fired at him and hit him. He jumped up and tried to run, but a second shot struck a vital spot, and he fell to rise no more. His skin was torn and gashed in many places, his nose was bleeding and he was a miserable looking object. The sod where they had been fighting was torn up in many places, and covered with blood and hair. As I had no sympathy with bears, it was quite an entertaining sight to witness such a fight.

Although I have heard of numerous cases where men have fought with and killed them with a sheath knife, when I attack one, I want a good rifle and a tree close at hand. Bears are very numerous all along the coast. They are not hunted much, as their skins are of but little value, and the hunting of them is attended with some danger. Although their disposition is to let you alone, if you let them alone, yet they will attack you if wounded. They will also fight bravely in defense of their young. Their flesh is of little value as food. They live on berries and roots and the salmon that die while ascending the creeks and rivers in the spawning season. The bear hunter must be a good shot, and possess a steady nerve. He must use a proper amount of caution, for, if careless, it is only a question of time, when instead of getting the bear, the bear will get him.

Chapter XIII

ONE WOULD NATURALLY SUPPOSE that in this high latitude the winters would be very cold, yet it is not the case. The so called Japanese ocean current keeps the climate warm here on the coast. I never saw the mercury in the thermometer lower than twenty degrees below zero, and then only for a few days. It is not so cold as in New York and Northern Pennsylvania. But not so the interior. The coast range of mountains is an effectual barrier to the influence of the ocean current, and on and east of the mountains the cold is very intense. I shall speak of this hereafter.

There are a great many glaciers here on the coast range of mountains. One of the most noted is the Muir glacier. It is a tremendous body of ice. This glacier empties its immense flow of ice into the so called Glacier Bay. It has a frontage on salt water of about three miles and a depth where the ice breaks off of at least three hundred feet. No one knows how far it extends back through the coast range, although it has been explored a distance of sixty miles or more. No end has been discovered. This vast body of ice is moving to the ocean at the rate of forty feet in twenty-four hours. It extends out over the sea until its unsupported mass is heavy enough to break itself off. Then it falls in large masses into the sea with a tremendous splash making waves which would swamp any ordinary small boat. These masses of broken ice are sometimes very large, and drifting to sea become dangerous to navigation.

The sound caused by this falling ice is nearly continuous, and at the distance of a mile or so is not unlike artillery firing. It is a grand sight and one never to be forgotten.

No intelligent being can look upon this scene, without feelings of reverence and admiration for the creator of all this wonderful grandeur. In it we recognize the works of the Almighty, the handiwork of Him who made and rules the planetary system so vast and far beyond our ability to comprehend, that we only see the beginning. And yet the All Powerful does not forget us. All His mighty works contribute to our welfare. Although our poor, weak minds, may not always be able to see how the benefit is to come from some things; all have their use, and nothing is made in vain. The glaciers along the coast are numbered by the hundreds. The large majority of them belong to a class called dead glaciers. They do not move nor flow. They are rapidly melting away. Every summer reduces their size. This must be due to a change in climate, and indicates that the climate is much warmer now than when those huge masses of ice were formed.

To travel on their surface is very dangerous as they are broken up by deep crevasses in which no bottom can be seen. A great many years ago two Russian officers in exploring the surface of one, disappeared and were never heard of afterwards.

Why should American tourists go to Europe for scenery? The mountain scenery of Alaska is unexcelled, and its glaciers are the grandest in the world. They are so pronounced by those who have visited Alaska and also Europe. For convenience of access they are unexcelled. One can take the steamer at Port Townsend, Washington, and see all the coast scenery without leaving its deck. The route from Puget Sound to Alaska is island bound almost its entire length. There are hundreds of islands both large and small, protecting the ship from the ocean swells, and making the trip a smooth and pleasant one. Every year the route is becoming more popular for tourists, and there is no doubt but that time will eventually make it one of the most fashionable resorts on earth.

Thus Alaska, although sparsely settled and little known, is destined to become eventually a great wealth producing country. It has mines of gold and also of silver, for lately several good silver prospects have been found near Juneau, and more are being continually discovered. Like all new mines, some of them will prove to be passable, others will not produce sufficient to meet expenses and cannot be worked.

Chapter XIV

IN ALL NEW MINING DISTRICTS, by far the greater number of locations made by the mining prospectors, upon thorough investigation prove to be worthless. Other ledges or deposits produce barely enough to pay running expenses, and a few prove to be immensely rich. Besides paying all expenses incurred for extracting the metals from the ore, they pay the owners large profits.

Then again there is a class of men in all new districts, who buy or locate claims in close proximity to the richer and more famous mines in the districts where they may chance to be. The real value of their claims is a matter of little consequence to them. The imaginary value of the claims is then represented by issuing stock. This is frequently sold on the merits of the rich mines adjoining. The wildcat schemers claiming that their ledges are the continuations of those already proven to be rich, and that they only require development to be equally productive.

Having sold large amounts of stock, they then proceed to assess the stockholders, on the plea of raising money for development, work, mining machinery, etc. Eventually the stockholders get tired of this continuous drain and an investigation shows that they have been swindled. There is no redress, and they are poorer, but wiser men. These swindlers are very injurious to legitimate mining. By floating worthless stock they destroy the confidence of the public, making it much more difficult for those, who have mines of real value, to raise funds to develop their mines and procure machinery to successfully operate with.

Another disadvantage has been the absence of civil law in the Territory. For years it was impossible to procure a perfect title to property, and mining men did not care to invest their money where there was no law and no protection for life or property. But fortunately now these difficulties are removed, and the operator can go ahead with a certainty of being protected in his rights.

One great advantage enjoyed by Alaska people is an abundance of water power. The heavy rainfall, glacier ice, and winter snows, make many streams of pure, fresh water, suitable for any purpose. The great Treadwell mill is run by water, although it is also furnished with steam power. This is sometimes required in winter, when the frost is severe enough to cause a scarcity of water. The abundance of water and cheapness of fuel enable the miner to work ores much cheaper than it can be done in any other part of the United States.

Living in Alaska is not so expensive as it is in Idaho or Colorado. This is due to cheap transportation by steamers from southern ports. Although they raise no vegetables worth mentioning, fish is very plentiful and costs but a trifle. The headquarters for miners' supplies — for almost the whole Territory — at present, is Juneau City. It is likely to continue so, being centrally situated and easy of access. Here, also, is the supply point for the Yukon miners; although, of late, the Northwest Trading Company have been shipping supplies direct from San Francisco to the mouth of the Yukon River. Here they again ship them by a smaller steamer up the river, meeting the miners near the head waters. This makes it much more convenient for miners to procure supplies than in former years. Then all supplies had to be packed from the coast by Indians to the head of navigation. For small boats on Lewis [Lewes] River the expense was much greater.

Chapter XV

HAVING PROSPECTED AND MINED on the coast for several years with varied success, I determined to try my luck in the interior. One bright April morning found me in readiness for the trip. I started with only two companions. Several other parties were going also, and we stayed together as long as it was convenient. We were obliged to take supplies enough to last us six months, as we intended to remain during the summer, and at that date no supplies were coming up the Yukon River. We employed Indians to take us to the head of Daya [Taiya] Inlet, which is the starting point to go overland. We arrived safely with our stores at this place. There is a portage to be made. It is thirty-five miles from the head of Daya Inlet to the fresh water lakes, that empty themselves into the Lewis River. Further down the Lewis River unites with the Pelley [Pelly] River, forming the Yukon.

We employed Indians to pack our supplies to the lakes, and were obliged to pay them about twelve dollars and fifty cents for one hundred pounds. The price was too much, but they had an understanding with each other not to pack for less, and we had to pay it or do our own packing. This we could not do, as our outfits, tools and all were near a thousand pounds to the man. Those Indians are known as the Chilkoot tribe. They are very saucy, and although not actually hostile they do not like white men, and are extremely impudent. They will not allow other tribes to come there to pack, as they claim the trail, so they fix their own prices, which are outrageous. Also their chiefs demand one dollar per head from every white man passing through their country. Very flattering to one's feelings to be obliged to pay a dirty, greasy, disgusting looking savage one dollar, demanded in a half threatening manner, for the privilege of passing through Uncle Sam's domain.

Yet this tribe is very powerful, and to quarrel with them would make the trip impossible. We controlled our passion for peace and profit's sake. The natives were very lazy, and it was a hard job to get them started. On some trifling pretext they would delay us from day to day, and it was ten days from the time we landed till we finally got off. Then the whole gang turned out and carried all our goods over on one trip, taking three days to make the trip, traveling near twelve miles per day. Some of the stronger ones carried very heavy loads, as much as two hundred pounds; others one hundred, and boys and squaws carried light loads.

Although there was a regular trail, the snow was still on the ground. In some places it was solid enough to bear us up, in other places it would break through every few yards. Then it would have to be beaten solid in order that the heavy packs might be carried over the trail. We went along the Chilcoot River* about twenty-five miles to the summit of the coast range of mountains, and at the summit passed between two high peaks, and it became very steep. As we ascended the snow became deeper, but firm and solid, so that the heaviest packers could walk safely on its surface. Near the summit it became so steep that it was necessary to cut a stair-case in the hard ice-like snow, in order to make it safe for those carrying burdens. After gaining the summit there is a comparatively level plateau, or vast snow plain, extending many miles on the summit of the range. Standing in the center of this great snow field you can look in any direction, but you will see nothing except snow and sky. Not a bush nor shrub, not even a blade of grass can you see to relieve the eye of the glaring white and azure blue that meets the view at every point. Here it is necessary to wear goggles of colored glass, or some colored crape-like material over the face to prevent snow blindness.

When we reached this field of snow it was midday, and the sun reflecting on the snow was almost unbearable. Although I had taken the precaution to partly cover my face, I was so severely sunburnt that a few days after the skin peeled off my entire face and the back of my hands. Traveling a few miles further we came to the edge of the plateau, and could see the lakes and the timber on their borders. Below we could see far down the valley of the Lewis River, as it wended its way northward, we knew not where.

*The Taiya River which heads in Chilkoot Pass.

From the edge of the plateau to the lakes it is about eight miles. The descent is quite steep in some places, but we accomplished it without accident. We paid off our Indians and camped on the banks of lake Linderman [Lindeman]. We afterward discovered, on opening our packs, that many of them had been opened by the natives, articles had been stolen and the packs again done up as we had done them at first. This was a source of much inconvenience to us. The things stolen were tools and articles that were almost indispensable to us; and we were in a land where they could not be replaced.

The lakes were still covered with ice, although it was now the tenth of May. The timber where we were was not very good for boat building. So we concluded to make sleds and haul our goods by hand on the ice to some point further down, where we could get good timber. It required one day to build our sleds, and early next morning we started with our first loads. We found good timber on lake Bennett, twenty miles below; and by making several trips, and taking advantage of the hardened ice in the early mornings, moved every thing there. We at once proceeded to saw lumber with whipsaws and build boats. We built the boats strong and large enough to carry all our supplies, made oars and got everything ready. Then we were obliged to wait two weeks for the ice to break up in the lakes before we could get through. At length the ice broke up; we embarked and worked our way northward through the masses of broken ice. We passed through lake Bennett to Mud lake, and found the outlet that formed the head of Lewis River. The country about those lakes consists of high mountains of bare rock. Around their bases are some small valleys and flats covered with scrub timber, but at a slight elevation there is no vegetation of any kind. It is a gray, barren, solid-looking rock. The rainfall here is very light, as the coast range of mountains appears to be the barrier to the rainfall and prevents the heavy rain clouds from the ocean from carrying their moisture inland. This accounts for the dry barrenness of the lake district.

The winters are very severe among those bleak, rocky mountains, as they are at a great elevation above sea level. There is but little game here; but the lakes are well supplied with fish. The different kinds in the lakes are white fish, pike and Arctic trout, and the rivers are well supplied with salmon in summer. The Arctic trout is a fish peculiar to the Arctic Slope. It is a fine flavored fish, and bites greedily at a fly hook. The white fish, pike and salmon are caught in nets and traps or speared.

In this lake region there is a small tribe of natives. They are very friendly to the whites, but will steal at every opportunity and are very dirty. They are miserable wretches and have miserable homes. Their clothing consists principally of the skins of rabbits; also of the poorer skins of the more valuable animals. They sell their furs to the coast Indians and get very little for them.

Chapter XVI

WE ENTERED THE OUTLET OF THE LAKE and drifted down with the current. This was a very pleasant mode of traveling, riding with the current in a good boat, only pulling oars when it became necessary to get steerage way. The sun was shining brightly, and just warm enough to make one lazy. The trees on the banks were leafing out and the birds gayly singing. There were several of the migratory species of birds that I recognized. The lark and robin redbreast, the whippoorwill and swallow.

The kinds of timber became more plentiful and larger as we drifted down. They consisted of spruce, hemlock, Norway pine, cottonwood, balm of Gilead, alder, willow and some other varieties of shrubbery, with which I was not acquainted.

We were obliged to keep a good lookout for rapids or

falls. This drifting down was really quite pleasant; but I could not avoid the thought that the drifting up part of the business on the return trip in the fall would not be so pleasant. The river was low, and we occasionally stopped at the mouths of incoming streams, and taking out a pick, shovel and gold pan would try the gravel to see if we could raise a color of gold. This we did in several places, and found that there was gold in the country. This was encouraging, and, as prospectors always have a feeling that it is richer ahead, we kept on.

We drifted along until the evening of our first day after entering the river, when we came to a canyon. Here the river contracted to a narrow gorge. The walls are perpendicular on both sides — about one hundred feet above the surface of the water. The canyon is about a mile in length. We camped on the river bank at the upper end of this canyon, and walking down its bank examined it carefully to see if we could run through with our boats. We saw that to run through would be attended with some danger, and that it would be impossible with our heavy loaded boats. The water runs through this canyon with tremendous velocity, and in places forms breakers which surge up and down so rough that a heavy laden boat would be sure to be swamped. Also it would be necessary to be very careful not to strike the wall rock; for the force of the boat running at such a rate would undoubtedly break the boat. We concluded to carry our stores over the hill to the lower end of the canyon, where we could land and take them on below, and then run through with the empty boat.

The next day, having carried our supplies down, we got in and pushed off. I must confess I felt a little shaky, and as we got into those tremendous boiling breakers I felt decidedly uneasy. At every plunge of the boat the water came rolling in over the sides. We were traveling at railway speed; we were half through and the boat was half full of water. But fortunately we did not take water so fast on the lower end of the run. Just as we made the shore below our boat filled to the top of the gunwale. We quickly jumped out, which, of course, lightened the boat and we were safe once more. The other boats had about the same experience, and one filled entirely. The boat sank only to the top of the gunwale, and the men clung to the top of the boat. We pushed out with our boat and pulled them in.

We reloaded our goods and started again. We only traveled about three miles until we came to a waterfall. There we had to unload and portage the boat and everything about one hundred yards. We handled the boats by laying down skids and drawing the boats over them.

All aboard again; and it was all plain sailing. There were some rapids, but no more falls on the river, and no more portages to make for a distance of more than two thousand miles, or to the river's mouth.

We floated lazily down the river, the climate becoming warmer and vegetation more rank and plentiful. We also began to see game, and finally drifted out of the river into a large lake of fresh water. This lake was about sixty miles long by twenty in width. Its banks, like those of the lakes above, are mountains of rock; bleak, gray and forbidding in appearance, vegetation growing only near the water's edge.

In this lake are several small, rocky islands. There geese, ducks and gulls had made their nests, and we got quite a lot of their eggs. They are not so palatable as hens' eggs, but will do very well in the absence of a better article. We passed through the lake and again entered the river. The weather remained pleasant; the days were warm and fine, and the nights cool, just right for sleeping.

We stopped occasionally to prospect and hunt, as we were in a game country. There were moose, caribou and mountain sheep. There was gold on the river bars, and some fair prospects; but the river was rising rapidly every day, covering up the level of the gold deposits. This was caused by the melting of snow in the mountains, and we knew that the water would not fall again for two months. We could not find any high bars, so we kept on down. We passed through a country that appeared to be rich in coal. A number of veins of this fuel cropped out from the river' banks, which appeared to be high mountains of clay of a blueish color. Coal was plenty here, but no gold.

Chapter XVII

IN THIS CLAY DEPOSIT WERE NUMEROUS BONES of animals that must have lived there ages ago. Among these bones we recognized what we supposed to be the bones of the mammoth, old tusks of ivory, cracked and black with age, also smaller, short, thick-looking tusks, that might have been the bones of a rhinoceros. Some of these were very large. Undoubtedly the bones were the bones of animals that were larger than any animals on the earth to-day. We passed on through this coal formation until we came to the mouth of the Pelley River. There the junction of the Lewis and Pelley Rivers forms the Yukon. We had passed the Hoody Link [Hootalinqua, now called the Teslin] River and the Big and Little Salmon Rivers higher up.

Now we were drifting on the Yukon. We came to a slate formation, indicative of precious metals, and again commenced prospecting for gold. We would land, and while some would hunt others would prospect. We found plenty of game, but little gold. We kept on drifting down, passed the mouth of White River and came to the Stuart [Stewart] River. We decided to go no further but ascend the Stuart River.

This we would be obliged to descend again before cold weather, for at that date there were no supplies in the country, except what the miner carried and some game, and the game could not be relied on. A district will be abounding with game one season, the next probably not a single animal can be found. The country is large, and the animals keep moving and getting away from places where they have been hunted. The Indians overcome this difficulty by following up the game, but a miner cannot do that and mine too.

Here we saw quite a lot of natives. They are a fine race, cleanly and honest, and an exception to the general Indian character. We were a curiosity to them, many of them having never seen a white man before. They were friendly and brought us venison. They are tall and straight, and have regular features. They are truthful and virtuous in every respect. They wanted to trade with us. We traded a little with them; but our stock of goods for trading purposes was light, as we came to mine not to trade. Their honesty and good behavior are attributed to the teachings of a missionary, who went there and taught them many years ago. This missionary went among them alone, before they had any opportunity to become debauched. His teachings have been the cause of their prosperity. I tried to find out the name of this noble minded man, that I might put it before the public, but I did not succeed, as the natives only called him father. We could make caches of provisions at any place in the territory inhabited by this tribe — only securing things against wild animals by making scaffoldings high up from the ground. On our return we would find all things safe, although we could see where they had camped at times right beside our caches.

Among those Indians we found a half-breed Russian who could talk a little English. He acted as interpreter. He told us he had come from a Russian settlement down near the mouth of the Yukon River. He had married one of their daughters and intended to remain among them. But this tribe is an exception, there being no other Indians like them in all Alaska.

There was only one thing in their mode of life that I did not like. They made slaves of their females, making them do all the heavy and menial labor. They are a migratory tribe. They follow the game in winter, but in summer they camp along the rivers and fish.

Their canoes are made of birch bark, sewed together with a small, tough, string-like root, which they get from the ground. These canoes are very light, and are made of different sizes. The larger ones will carry about a ton, the small ones one man. They are very neatly constructed and quite handsome.

Their clothing in summer consists of the tanned skins of the caribou and moose. In winter they dress in the furs of the fox, lynx, otter, marten, beaver and bear.

Having lightened our boats by caching some of our provisions, we commenced ascending the Stuart River. The water was high, and rising every day. It made the work much harder, as the current was stronger and we were obliged to keep close in shore, so that we could reach bottom with our poles. Some times we would take a line ashore and tow, one man remaining in the boat to steer. It was hard and slow work.

The Stuart River valley is, in places, fifteen or more miles in width, and again the mountains will close it in to two or three miles. The valley is covered with timber, the varieties being the same as I have previously described. The mountains bordering this valley are not high. They are covered with moss and short grass, and in places small groves of shrubbery. In ascending the river we killed several moose — a welcome addition to our larder. The flesh of the moose is almost, if not fully, equal to beef. A full grown, fat moose will make eight or nine hundred pounds of excellent meat.

In many places the high water was cutting in under the river banks, which were principally of sand. The sand was frozen, for in this high latitude the ground is frozen all the year — no one knows how deep. The miners have sunk holes in search of gold thirty feet in depth and it was still frozen. In the summer it thaws out two or three feet at the top, but freezes again in the winter. The river water has a temperature much warmer than the frozen river banks, and the current rushing against the banks thaws out and undermines great blocks of it. It often washes away groves of timber. In this manner those trees and blocks of frozen surface fall into the river, giving but little warning.

And now the reader can plainly see wherein lies the danger in ascending the stream. We had some narrow escapes, and one party lost their entire outfit, boat and all, by a fall of land and timber. Fortunately no one was hurt. We were then obliged to divide our stores with them and assist them to build another boat. In this manner we toiled and tugged, pulled and poled, but had so many difficulties to contend with that our progress was slow. However we kept at it late and early, and by the fourth of July we had ascended about one hundred miles. The river had commenced falling, and from that time on lowered rapidly. As the river bars became dry we found some gold. We built rockers and went to work rocking. The result was twenty-five dollars per day to the man. We kept on rocking. We also prospected along the side streams in the vicinity, but found nothing of any value. The gold on the river bars was very fine and limited in extent.

Chapter XVIII

TWO-THIRDS OF OUR PARTY had weakened and turned back, leaving but seven of us. As the bar diggings were not satisfying us we decided to give our attention entirely to prospecting. We went some further up the river, prospecting as we went along, also prospecting side streams. We found coarse gold in several places; but it was slow work to sink holes, the ground being frozen so hard. We also took trips on foot through the surrounding country.

The climate here in summer is very dry, and the days are even hot. From the middle of June until the middle of July there is no night. I plainly saw the sun shining on the mountain at twelve o'clock at night. It is very proper that it should be so, for this is called the land of the midnight sun.

This is a good fur country; reasonably good for moose, and as usual, more bear than we wanted. One morning a companion and myself went to a small stream that emptied into the river near our camp to prospect. We had no arms with us — nothing but a pick, shovel, and gold pan. I was carrying the shovel and pan and walking ahead. We came to an old game trail that would apparently lead us up the creek, so we followed it. In places it was covered with brush and partly hidden. As we were passing through one of those places a monstrous brown bear rose up suddenly in the trail. He was so close that I could almost touch him, and I was very much afraid that he would touch me. He stood upright on his hind feet in a kind of John L. Sullivan position, and snorted at me. I was afraid to turn for fear he would attack me behind; and I think the bear had the same feeling toward me. I had no arms; not even a knife. Taking my gold pan in one hand and my shovel in the other, I made music for him. He snorted again and backed off a few steps slowly, and finally clearing the trail with a bound he ran away at a terrible pace. I hardly know who was scared the most, the bear or myself.

Which is the worst scared?

FROM THE ORIGINAL EDITION

Although we had done considerable prospecting we found nothing to repay our time and labor. Having a curiosity to get a bird's eye view of the country, I took a trip one day alone to the summit of the highest peak that I could see in that locality. Expecting to return the same day I only carried a small piece of bread with me for lunch. Before I arrived at the base of the mountain I found that it was fully twice the distance from camp that I had calculated on. I commenced the ascent, which was not steep at starting, but gradually became so as I climbed. The mountain was also higher than I had anticipated. It was four o'clock when I sat down to rest on its summit.

A kind of short, hard grass grew on its slopes, but in places the grass was replaced with moss. I believe it was the kind that the reindeer feeds on. There were also clumps of shrubbery and a great many kinds of wild flowers; some of them very beautiful. The view from the summit was grand in its immense extent, but not inspiring. It was lonesome in the extreme. As far as the eye could see in every direction, it was a vast solitude. Not a sound could I hear on the summit of that mountain; not even the twittering of a bird, nor the hum of a bee. The animal life that exists in this country appears to live only in the valleys, and the country is so cold that nothing can live in winter.

But little food is produced in this far northern region, so that no great amount of birds and beasts can find support here. Excepting the shrubs and grass that feeds the moose and reindeer there is little feed for anything, and the moose and caribou, or reindeer, confine themselves to the river bottoms. The rolling hill country is a land of desolation. Nothing breaks the stillness of this vast territory; no animal nor bird in sight, and not a sound to be heard. I cannot describe the feeling it creates — one of extreme loneliness in one who witnesses it. Although it was getting late, about six o'clock — as the days then were twenty-four hours long — I was not afraid of being caught in the dark. I had noticed from the summit a gulch that lay to the eastward. I decided to return that way. We were short of fresh provisions, and as there was timber in the gulch I thought I might get a moose or caribou.

To reach the gulch it was necessary to descend farther eastward than where I came up. It did not take long to get down, and I was rewarded by seeing game signs. Passing along quietly through the timber, I found some fresh tracks and followed them.

I had not gone far when I saw two caribou, a cow and her calf, feeding in a clump of willow brush. Aiming at the cow I fired and killed her where she stood. The calf became frightened and ran. I fired and brought it down, but it jumped up and ran again. I knew it would not go far, so drawing my knife I bled the cow and then followed the calf. It traveled faster and farther than I thought it would. However I found, shot, and quartering it, hung it up in a tree to keep it safe from bear or lynx. This took some time. I had been gone from the cow about three hours. I felt quite hungry, so I made a fire and roasted some of the meat. Having satisfied my hunger I went back to the cow, which was more than two miles distant. When I got near I heard growling. I knew that bears had smelt the blood and were eating my caribou. Crawling quietly up I saw a large cinnamon bear and two half grown cubs. They were having a fine feast at my expense. I looked for a tree to climb in case my shot was not fatal. I did not fear the cubs, but I knew the mother would fight to the last.

There was no suitable tree standing near, but a tree had fallen, lodging on a high rock. If close pressed, by running up this tree I could gain the rock; and the only way to gain it was by the tree trunk. So getting as close as I could safely I fired. She fell, rolled over, and jumping up stood up and clawed the air. Another shot and the beast went down again. This time she saw me when I fired, and although mortally wounded, jumped up and came rushing furiously at me. My rifle was a Winchester of the most approved pattern. Pumping from the magazine I fired rapidly, planting every shot in the big brute's body, some times knocking it down and at others only checking her progress. It was quite a noisy affair, the old bear roaring with pain and the cubs growling and snarling. The bear kept coming and I kept firing. She was getting too close for comfort. I had pumped and fired my last shot from the magazine. As I was wearing moccasins I was sure footed. Running to the log, which was near by, I leaped upon it and gained the rock. The brute came to the log. I was replenishing my magazine, which I quickly did, and was just going to fire again when she stopped. She was bleeding from a number of wounds, and quite a stream was running from her mouth. Her fore feet were on the log and she was swinging to and fro. She had quit growling, when suddenly a great crimson stream poured from her mouth and she fell dead. I then turned my attention to the cubs. They had run up and were howling and smelling of the body. A few shots quieted them. They had spoiled quite a lot of the caribou. I cut up what was good, hung it up and went to camp to get help to pack in the game.

Chapter XIX

WE WERE NOW BEGINNING TO THINK of our return. They had all been gone for a month, except my party of seven. As we would be much lighter on the return, not carrying tools, nor so much provision, we decided to build two lighter boats, one for three and one for four. The timber was good where we were camped, so we commenced building at once, and in a few days completed two strong, light and well shaped crafts.

We embarked and went to the mouth of the river, where we had cached the supplies for our return, only to find the most of the provisions destroyed by fire. The fire had got out from the camp-fires of the men who had gone before, and had destroyed fully two-thirds of the provisions and badly damaged the remainder. The situation was unpleasant. It was eight hundred miles to the nearest supplies, the traveling being mostly up the river. We had a strong current to contend with, but having a good supply of dried caribou we thought we could make the trip. We worked hard every day, starting early, pulling, poling and towing; only stopping when tired and in need of rest and refreshments.

We were now beginning to have nights, as the nights in

winter in that land increase in length, until like the day in summer, it is all night. The river was almost as high as when we were drifting down, which I afterwards found was caused by heavy rains in the coast range. This was unfortunate for us, as the current is much stronger when the river is high. The parties ahead of us had done us a great deal of injury in being so careless with their camp-fires. They had not only burnt our provisions, but had allowed their fires to get out all the way up the river. This had scared away the game; and where there was game when we came down, and we expected to get some when going back, there was not a track to be seen. We caught some fish, which was a great help to us; but we could not fish and travel at the same time. Only one variety would bite a hook, the Arctic trout, and they being small it required a great deal of fishing to amount to anything.

When we got to Pelley River we found our provisions were going so fast that unless we got some game we should die of starvation before reaching the coast. We decided to stop and hunt one day. We scattered out in different directions, but the big game had all left. We got nothing but a few fool hens, a kind of grouse, and two small rabbits. We then decided to put ourselves on rations, which we did, and also traveled from sunrise to sunset.

The days were yet twenty hours long, but were shortening very rapidly. We found we could not stand this, so we shortened our working hours. We also worked the smaller boat with two men and the larger with three, and kept a hunter out on each side of the river to pick up anything that could be used for food. In this manner we accomplished five hundred miles. There were three hundred more before we could expect relief. Our provisions had entirely given out, with the exception of a little tea and a small sack of salt. The weather was also growing cold. It was now late in September, and in this high latitude the rivers freeze in October. A miserable situation — no food, no tobacco, our clothing almost worn out. Every night appeared to be colder than the preceding one, yet we kept ahead, determined to do our best. Sometimes we would get a rabbit or a grouse, which we would carefully divide. We were getting weaker every day, and yet all of us were kind and pleasant to each other. Sometimes we would hear a complaint against the careless parties, who, having first destroyed our provisions, through further careless actions had driven the game from our route.

We could still, at times, catch a few fish, but the further up we got the harder they were to catch. It made our hands very cold to handle the wet oars and poles in the boat. I cut a piece of blanket and made a pair of mittens. All the rest followed suit. They were a great benefit. Although they would get wet they still kept our hands warm. We accomplished another hundred miles.

Here some of the men wanted to leave the boats and travel afoot to the coast. To this I strongly objected. I argued with them a long time, pointing out the danger attending the attempt. The mountains were covered with snow; it was getting deeper every day. The cold was increasing, and the country they would be obliged to travel through was an unknown wilderness, of brush, swamps and mountains. There would be swift mountain streams to cross. In their weakened condition they could not carry much bedding, while, if they stayed with the boat, they at least had a good bed. But all to no purpose — I argued in vain. Three were determined to go. The other three believed as I did, and stayed with the boat. They left us the following morning, taking with them a shot gun, one rifle, and a little salt. Each one carried a blanket. We parted with them sorrowfully, for I felt almost certain that they would never get through. The poor fellows have never been heard of since. There were but four of us now, so we only took one boat. I can hardly tell how we got along. We traveled all our meager strength would allow, intending to stick to the boat, and if the river froze up then we would travel afoot. One day as we were boating along I saw something swimming near the opposite shore. It had started to cross, not having seen us. We kept quiet, and it came close to us. It proved to be a lynx, a large, fat fellow. We killed him and had a feast at once, with meat enough left for several days. At length we came to the canyon.

As I spoke of passing it on our downward trip, it is not necessary to describe it now. Here we were obliged to portage the boat a distance of near one mile. Weak as we were from hunger, this was a hard job. We traveled from the canyon to the first lake in two days. Now there would be at least no more current to contend with. We were on the lakes, but the wind was blowing, and continued to blow so hard we could not travel. We had not eaten any thing whatever for three days, and but little before since we got the lynx. I took my rifle and started for the timber, though so weak I could hardly walk. One of the others also went hunting and the other two lay down. I walked around slowly through the timber. At times I would get so faint that I would have to sit down. What a terrible craving and gnawing at the pit of my stomach. I could not see even a little bird. I traveled slowly on through the timber. Sometimes I imagined that I saw game, and once I fired at a supposed rabbit. The rabbit did not move. I fired again. No move. I went up closer and saw that I had been deceived. It was a root; and as it was not an eatable root I quit firing at it. I kept on hunting, all to no purpose. There was no game to find. I came to an old, dry, half rotten tree. I started a fire, and laying down some bark lay down to rest by the fire. I dreamed that I was in a large city. I was looking at the window of a restaurant or bakery, and could see piles of bread, pies and cakes. I went in and sat down to a well furnished table and was eating. How long I was eating I did not know, but I could not satisfy myself. It did not do me any good. The more I ate the more I wanted to eat. When I awoke it was dark.

The fire in the old tree had burned low and I was quite chilly. I did not think I could find my way to camp in the dark, so I replenished the fire and sat there studying on what it was best to do. I knew that if we did not get something soon to strengthen us, some, if not all, would die.

All at once an idea came to me. I could hardly wait till daylight. As soon as gray dawn appeared I went back to camp. Bringing out three fox skins that I had traded for with Indians, I picked and singed the fur off. They were valuable skins. One of them, that of a black fox, was a rare specimen, and was valued at one hundred and fifty dollars. The other two were silver gray, also valuable specimens worth seventy-five dollars apiece. The total value of the three skins was three hundred dollars. It was all the same. Had they been worth three thousand, instead of three hundred, they must go into the pot after being picked, singed and scraped until clean.

I scalded them in hot water and scraped them. I then cut them fine and boiled them quite a while, putting in some salt. After a good deal of cooking they became quite tender, and tasted much better than I thought they would.

While I was doing this the others were yet in bed. I called them out and we divided the stew amongst us. We were strengthened at least for a few days more. The boys often afterwards laughed about the inglorious end of those high toned fox skins. I sometimes answered with the assertion, that had it not been for my fox skin stew they would still be sleeping on the shores of lake Bennett.

The wind calmed down, we launched the boat and traveled two days. Our stew had given out and we were hungry again. I was requested by the others to walk along the shore in quest of game, as I was the most successful. We had only one shot gun. The rifles were not fit to use in killing small game; and the game was small indeed. Sometimes a rabbit or grouse would fortunately come in my path, but generally nothing but little red squirrels and little birds. But the little birds and squirrels were all very scarce; and although I was the strongest of the party, I was too weak from hunger to get over much ground. However I did my best.

We were getting nearer the coast every day, and I began to feel encouraged and so did the others. I once brought in a few mice, although I could not eat them. I thought some of the others might want to try them. They did, and called them good.

The weather became several degrees colder, the lakes froze over, and boat traveling was at an end. However we were close to the portage. The cold was terrible on us. We had to be careful not to freeze our hands and feet. The cold wind blew fiercely through our worn and ragged clothing. We staggered wearily along the lake shore for three days and reached the head of the lakes, and the end of the portage trail.

Now if we could reach the summit of the portage the balance of the trip to the coast would be down grade; only thirty-five miles further and we would be safe. But the first ten or twelve miles were all hard climbing. Again we were entirely destitute. Not a morsel of food. We were afraid to start in our starved condition. We decided to hunt one day and try to get some kind of a meal before starting. All four of us went out. We shot some small squirrels which did not do us much good. During the day myself and one of the others had gone into the timber up a small stream; he on one side of the stream and I on the other. Suddenly I heard a cry like the voice of a child crying. I listened and heard it again. I called to my partner, who came across to where I was. Again we heard the cry, this time louder than before. It was the cry of a lynx. We quickly went in the direction of the sound. The sound continued, and we traveled as fast as possible for fear that the animal might move and we should lose him. We came upon them (there were two of them, mates) in traveling about a mile. On seeing us they ran up a tree. A shot apiece settled them, and we went back to camp joyfully, loaded with meat. But before we went we made a fire and satisfied our hunger, or partly satisfied it; for we had been starved so long that we did not dare to eat all we wanted to.

Having returned to camp we rested and fed up two days, and then felt strong enough to climb the mountain. We filled our pockets with cold roasted meat and started. Although we suffered considerably from cold, we got through without accident. We found a newly established trading post on the coast and our troubles were over.

Chapter XX

WE THEN INQUIRED FOR THE MEN who had left us far below on the river. There were no tidings of them. If they had got through at all here would most likely be the point at which they would arrive. The snow was falling every day in the mountains, yet I was determined to make an effort to assist them. Only one of my companions would go with me. We replenished our clothing, dressing warmly for the trip — partly in furs. I hired two stout Indians and packed them with light loads of provisions, a couple of bottles of brandy, potted extract of beef, and such other things as would be needed. We started for the Chilcoot [Chilkat] Pass, thinking that probably we might find them in that direction. This was the pass through which they intended to come.* We traveled on an Indian trail, although it was mostly covered with snow.We traveled several days and came to the pass, near one hundred miles from the coast. The Indians stopped and would go no further. I wanted to go on through the pass. They would not go. Their wages were to be two dollars per day. I offered to give them five dollars per day if they would keep on. They would not go. I knew we could not go without them, for we could not carry enough to be of any benefit to the men if we did find them. I then told them if they would only go to the north side of the pass, where I could see down the north slope, I would then return and give them five dollars per day. Also if we found the men, and they helped us all they could to get them out, I would give them the same pay all the way back and make them a present of ten good blankets. The Indians like blankets. They talked a while by themselves, but finally said the pass was dangerous; that it was a glacier full of deep crevasses; that the crevasses were now hid by the falling snow, and we should lose our lives if we attempted it. They would not go. We had to give it up and leave the poor fellows to their fate, whatever that may have been. I got a long pole and finding a high knoll of loose rocks in a conspicuous place, fastened a red blanket firmly to the pole, and planted the pole firmly in the rocks. We then took out supplies just enough to carry us back and cached the balance at the base of the pole, covering it with rocks to secure it against animals. I then scraped the bark clean from the pole and wrote directions on it. This signal and cache were right in the path and they could not get by without seeing them. That was all we could do.

We returned to salt water, hired Indians to take us to Juneau in a canoe, and settled down for the winter.

No traces have ever been found of those three men, although years have come and gone since.

---

*This seems to have been the pass at the head of the Chilkat River, leading to Kusawa Lake.

Chapter XXI

SUCH IS THE LIFE OF THE FRONTIERSMAN. He generally starts early in life — a mere boy — and often from an eastern home. His first experience will be in some of the near territories, where, if he engages in mining and is successful, he may return to his home with wealth, and, perhaps, settle down as a farmer or business man. He sees men making fortunes and pursuing this course; he thinks he will acquire one also. He is only waiting for a stroke of fortune. Years pass by and he does not succeed in getting it. He becomes a prospector for precious metals. He is unfortunate and does not strike any thing of value. He goes farther west, thinking that in a new country his luck may change. More years pass by, and he is still a poor prospector. His hopes are aroused by every new excitement. All in vain. Fortune will not smile upon him. Perhaps he succeeds in accumulating a small fortune. He returns home with that, thinking he will settle down and be contented, but he is disappointed. The friends of his youth have gone, or changed. They are no longer boys and girls, but men and women with families. Many of his old associates are roaming the west, as he did, in search of fortune, and the old home is not what it used to be. His old familiar land marks only bring up sad memories of the past. He is lonesome and unhappy, and he again returns to the frontier; this time with no intention of ever returning. He lives his life out on the frontier, braving every danger for the sake of excitement. And when his end comes he is rudely buried by honest hands; often in a grave that is never again visited by a white man. No tombstone marks his resting place; no flowers grow upon his grave. But when the end of time comes he will be counted as good, or better, than those who have amassed millions by schemes which rob the poor and fill the world with misery.

On the coast I saw some of the parties who had preceded

us in coming from the interior, and who, through their criminal carelessness, had destroyed our provisions and driven off the game by allowing their camp-fires to get out in all directions. I could not refrain from telling them of the consequences of their careless conduct. I considered them to blame not only for our sufferings and losses, but I also believed them to be the cause of the loss of our three unfortunate companions. Some of them spoke of it with feelings of regret, but others only laughed brutishly. But those who laughed were not miners. They were creatures from the slums of cities whose modes of life had hardened their feelings. Such a man as this is not the kind of a man who will stand up bravely in the days of danger and hardship. He is the first to give up; the first to become a burden to his less pretentious companions. You can hear him with his feet elevated near a bar-room stove, bombastic and loud mouthed, recounting his deeds of bravery. But he must stop sometimes to take another dose of rot-gut whiskey to renew his courage; for here is where his deeds of bravery are done, at least all he ever does.

He sometimes mingles in a bar-room fight. His courage there is great, and especially so if he can find a small, weak or helplessly intoxicated man, with whom he can pick a quarrel. He will beat, kick and abuse him beyond description. He is so brave that, if others do not interfere, he may beat the poor, helpless creature to death. He has done a brave thing, and why not? Has he not been drinking whiskey all day? And if whiskey is not good to raise the courage in a ruffian I would like to know what it is good for. There is only one improvement he could make that would make him a still greater hero. He might take another drink and then go home and whip his wife or his mother. Of this class of men are those who laughed when I informed them of the consequences of their criminal neglect.

Now what is such a man in the presence of hardship or real danger? If there is hard labor to do he feigns sickness, and remains sick until it is done. If danger from crossing rivers he must have the safest place. If danger from Indians he brings up the rear, unless he fears an attack from that direction. Then he gets in the middle, and, if any real fighting occurs, he is so cowardly that he cannot fight. If food is scarce he must have his full supply. No self-denial there; no sympathy for an unfortunate comrade. If a fellow miner meets with an accident and is crippled, he will not assist. If a comrade loses his provisions this creature will spare him nothing, although others in the party will divide. But enough of such characters.

Now who is the man who does the work? Who meets and overcomes the dangers? He is the quiet, unassuming man. He never talks of his great deeds. In fact he hardly appears to know that he has done any thing extraordinary. If you want to find him do not look for him in the bar room. If he is not busy mining or prospecting then look in the workshop; or, if it is Sunday, go to church. There is where you will meet the right kind of a man. He also attends the meetings of temperance societies and literary societies. These are the men to choose for companions on hard and dangerous trips. They can be relied on. If you are sick they will do all they can to assist you. In the face of danger they do not shirk, but meet every thing bravely, quietly and with determination. They do not exult over a fallen foe. It is sad to think that we are ever obliged to take the life of a human being, even if it is only a low, degraded savage; but we sometimes must in defense of our own lives.

These men are not all members of the church, although some are, but all are men of good morals. If there is labor to be done all are willing and prompt. If food is scarce all are self-denying. They are on the best of terms with each other. There is no profane language, nor coarse, brutal jokes. In the evenings the conversation around the camp-fire is pleasant and interesting. And although gruff in their manners, they are honest. Many of them remember their Creator, and return thanks to Him nightly on retiring, for His never ceasing mercies. These are the men who explore the frontiers and pave the way for the future greatness of the country.

Although the large majority of frontiersmen are such as I have just described, there are enough of the other class to do much harm. They debauch and antagonize the natives, thus causing an almost continual warfare wherever the two races meet. God have mercy on them, for they have much to answer for.

Chapter XXII

I PASSED AN UNEVENTFUL WINTER AT JUNEAU and vicinity, part of the time in the employ of the Alaska Mill & Mining Company.* I was not satisfied with my interior trip; or, in other words, I wanted to return and try again. I believed and believe yet that there is gold in plenty to be discovered there, if one is only fortunate enough to find it. The gold-bearing parts of that country are very large. In fact there is a little gold in almost any place you may try. It was fine, or what miners term river bar gold. Occasionally a little of the coarse order is found in the small streams. But the ground being eternally frozen makes it hard to sink to bed rock; and it is generally on bed rock that coarse gold is found in paying quantities.

The summers are so short, that to go there from the coast and return the same season, allows but little time to work. So I concluded to go there again, and going in the early spring take supplies sufficient to last eighteen months. This would give me two summers to prospect, and I should have to pass one winter there. So being acquainted with some men who were going to stay the same length of time, and men who I thought would suit me as companions, I joined them. We procured our outfit, taking plenty of warm clothing and blankets, for we knew the winter would be cold.

We started about the first of March. We took small, but strong, hand-sleds with us, for we intended, if possible, to haul our supplies on the snow and ice to the lower end of the ice on the lakes. We knew the ice in the river broke up much earlier than it did on the lakes, and if we could get by the lakes to the river before the ice became too weak it would save us a good deal of time.

---

*This was the company that then operated the mines at Treadwell on Douglas Island.

We employed Indians with canoes to take us and our stores to the head of Daya [Taiya] Inlet. We arrived in good condition, and were again ready for the trail. The natives came and wanted to pack for us, but we remembered how they stole from us the year before, and would not give them any regular work. We sometimes hired them to haul sleds when we could watch them. We hauled our goods by making trips of one day's travel ahead and making a camp. Then we would leave some one to guard it, taking several days to haul up the remainder. When all was hauled up we would make another camp ahead, and in this manner prevented the natives from stealing.

Our progress was slow, but we kept steady at work, and had crossed the mountains and slid half way down the lakes when the ice became so weak that we could not reach the river. We camped and built boats and waited for the ice to break up. The ice broke up, we embarked, went to the canyon, packed our stores across the portage and ran the boats through without accident. We reloaded the goods, ran to the falls below, portaged them and then the river was clear of obstacles as far down as we chose to go. We drifted down with the current, as we had done the year before, letting the oars rest. The weather being warm and pleasant we enjoyed the easy, gliding motion of the boat, and smoked and talked, a very pleasant way of traveling. Down, down we went, past the mouth of the Hoody Link [Hootalinqua, now called the Teslin] River.

Chapter XXIII

HERE WE FOUND A BAR which we prospected and found sufficient gold to pay. Some of us stopped here and went to work, others kept on down. The water being very much lower than it was the previous year, we found a good many spots that would pay to work. When we found a place that would pay we would stop and work it out, or work, until the water, which was now rising, would drive us out. Then we would go on to another.

We kept on in this manner until the water had covered all the low bars. The higher bars were not rich enough to pay. We then gave our attention to the smaller streams emptying into the river in hopes of finding coarse gold. We would stop at every favorable looking stream, some times staying a week in one place.

I did not intend to go to the Stuart River this year, but intended to go further down the Yukon if the formation was favorable for gold. We kept working down gradually, and passed the mouths of the Stuart and White Rivers.

Sometimes we would take a trip back from the river forty or fifty miles, generally going in pairs. Our outfit on these little trips would be a pick, pan and shovel, a Winchester rifle, Colt's revolver, one-half a blanket and a little flour, bacon, sugar and coffee. We generally took very little of these things, and depended largely on game for food. In fact if game signs were plenty, I have started, intending to be gone several days, taking nothing to eat but a little salt.

Usually we would find a little gold at every place and not much at any place. A little adventure occurred to me on one of these trips, which may be interesting to the reader. This time I was alone. We only went hunting when we were short of fresh meat. I did not intend to go far, and took nothing but my rifle and hunting knife. We were camped on a flat place on the river bank. The flat extended about five miles back to the mountains. Our camp was at the mouth of a good sized creek, which was navigable for boats to the foot hills. The water being sluggish and deep I knew that my best chance for game at that season would be along the base of the hills. I followed up the stream, thinking that if I got any large game we could take a boat up the stream and thus save a good deal of packing.

I followed along the stream and as I neared the foot hills saw signs of moose. I was then at the base of the hills. Skirting along the hills a mile or more I came to an open prairie or glade, probably containing four or five acres. I sat down on a fallen tree to rest and look around. I had not sat there long when a large moose came out of the timber on the opposite side, and was coming directly towards me. I at once concealed myself behind a tree and kept quiet. He came on, and came so close to me that I could have easily hit him with a stone. I shot him behind the fore leg and he fell dead in his tracks. He was a fine specimen of his kind. I at once proceeded to cut him up, hanging the meat up in a tree. I then made a fire and cooked some of the meat, roasting it on a stick.

Having finished my dinner I thought I would climb up the mountain and get a good view of the country. It was yet early in the day, and I should still have plenty of time to go to camp and come back with help in the boat for the moose. I commenced the ascent. I saw plenty of fresh game tracks, but did not wish to kill any more, because the moose would be all we could use before it would spoil. I did not follow them, but kept on up the mountain. I reached the summit. It was not very high, but afforded a good view of the surrounding country.

Looking to the westward I saw a large lake. I could not see the length of it, but it appeared to be three or four miles wide. I could see that it was connected with the river through a break in the mountains below. But how much of it was lake and how much river I could not tell. It extended up to the base of the mountain on which I was standing. I was about to return to camp when I heard a sound as if some one was chopping. I listened. The sound continued. Looking directly down at the part of the lake nearest to me, where the sound appeared to come from, I distinctly saw smoke.

Who could it be in that land of solitude and desolation? At first I thought it was natives. Then I thought of a boat that had preceded us down the river a few days. It must be them. I thought they had gone down; had come to the stream that drained the lake, and ascending it with their boat had got into the lake. It was not more than four miles to the smoke. I decided to go down and see them, and see if they had made any discoveries. I started rapidly down the slope, and soon came to the camp, when I was quickly undeceived. It was a village of natives. I should have liked to turn back then for fear of treachery, for I never could trust Indians. But it was too late, I had been discovered. The dogs came rushing and barking at me furiously. There must have been at least forty of them. They were very large and powerful brutes, and wanted to eat me without stopping to argue the case at all. Some children, who had been playing behind the village, ran screaming and crying back to their huts. Some women who saw me ran away howling. The dogs got worse. I got my back against a tree, for they were very threatening, and were really more like wolves than dogs. At length one, who was bolder than the others, sprang squarely at my throat. I struck him with the barrel of my rifle, but he came again. Leveling my rifle I shot the brute dead. This scared the others and they backed off.

Up to this time I had not seen any men. Now I saw them coming; too many of them. Most of them were armed. They had spears, axes, clubs and old flint-lock guns. I saw that I had got into a hornet's nest. To run was worse than useless, for the dogs would tear me to pieces. I kept my back to the tree and determined to make every shot in my Winchester do its duty. Yet I would not fire the first shot, but leveled my rifle — a hint to them that if they fired the first they would get the next.

What a din! Dogs barking, women and children screaming, the men much excited and talking very loud. Just then I saw running up from behind our friend, the half-breed Russian interpreter, whose acquaintance I had made the year before. He quieted the men, and they called off the dogs. The squaws and young ones stopped their music. The half-breed told me that they were a band of Indians who lived away back; that he met them there every year to trade shot, powder, tobacco, matches and such things for their furs, and that but few of them had ever seen a white man. I had no trouble in making peace with them. The Indian wanted pay for his dog, which, of course, I agreed to.

It was too late to go back to camp that night, so I accepted the hospitality of the half-breed trader. The Indian whose dog I had shot would not take money in payment. He wanted a blanket. To this I consented, and next morning went back to camp, taking three of the Indians with me. I gave the Indian his blanket, and the other two some trifling present. They went off well satisfied.

We took the boat and got the moose, and embarked to go further down the river, prospecting along every stream we came to. We at length came to a river which is now called Forty Mile River. It is a stream of considerable size. Here we found coarse gold, and got encouraging prospects. We made preparations to ascend this river. We were obliged to build smaller and lighter boats. The boats we used coming down being too cumbersome. The current of this stream was very swift, and the channel rocky and dangerous.

Chapter XXIV

OUR BOATS WERE SOON FINISHED and the trip commenced. As we ascended we found spots that would pay, yet they were small, and a few days would work any of them out. We had left much the greater part of our supplies at the mouth of the river. The work of ascending was thus much lightened. In places there were rocky rapids, some times even making short portages necessary.

At one of these rapids we had an accident, which was quite a loss to us, both of time and provisions. We had a tow line ashore with two men, myself and one other, towing the boat, and one man in the boat to steer. There was a large rock that lay in a position which made it very inconvenient to pass. From the position in which it lay it was necessary to pass it on the outside. We had got the boat by and were beginning to haul in when suddenly the boat took a sheer and strained so heavy on the rope as to break it. In an instant it swang broadside to the current, and drifted rapidly down stream. The steersman did all he could, and succeeded in bringing the boat near the shore, but it was in a bad place. There was the trunk of a tree leaning very low over the water, which was very swift and deep. The lower gunwale of the boat caught the tree broadside, and in an instant it capsized. The steersman got hold of a brush and hauled himself out.

Another party a short distance below picked up the boat and our bedding which floated, but we lost all the provisions and tools we had with us. Fortunately we did not lose our arms, as we had them tied under the forecastle in the bow of the boat, (a precaution which we always took) and they could not fall out. We got our boat and at once returned to the mouth of the river for supplies. Having obtained them, we returned up the river, and again started to go further up, working in places as we went. At length we came to a place where the river forked.

One fork carried no gold. We kept following up the other until the gold became scarce, and we no longer found spots sufficiently rich to work. At length we emerged into a low flat country, and by climbing a neighboring hill we could see many lakes and swamps. There was a range of snow capped mountains to the westward.

The country was covered with timber and swarming with mosquitoes. Game was scarce. This I attributed to these pests. There were so many of them, and they were so ravenous, that any other animal life could hardly exist. The sun now shone almost day and night, it being mid-summer. The gold on the river had caused the miners to concentrate there, and when we reached the flat lake country we found the rest of our party. Our party consisted of thirty-two men.

We decided to make our way through the swamps and lakes and try to reach the mountains to the west. We worked hard from day to day pulling through the lakes, and in places passing through narrow channels. Sometimes we were impeded by timber that had fallen across the channel. We then had to cut the timber out. In other places the lakes would be more swamp than lake. It would be a black sticky mud, through which it was hard to pull the boats.

We had a continual fight with the mosquitoes. They would fill our eyes, nose and mouth, and when we got through our hands and faces were covered with sores, caused by those poisonous insects. After ten days of hard labor we reached the mountains.

At their base were banks of clay, miles of it, on which grew no vegetation of any kind. In walking over it I found a great many bones, generally partly buried in the clay, and having the appearance of having been entirely buried at some time. And oh! what bones! I had no means of measuring them, but they certainly had belonged to animals that were much larger than any existing at the present time. What great, powerful and clumsy beasts they must have been. We also found teeth. In looking at them one could plainly see that they were teeth, yet they were so large that one could hardly realize the fact. The ivory tusks were very long and curved when not broken; yet it was hard to find one that had not been broken. They were so cracked and black with age as to be entirely useless for anything except relics of past ages.

How or why did those animals become extinct? Was it owing to a change of climate? Indications point that way. If those old bones could speak, they might tell us great things of doings in ancient times, that we do not now even suspect. But bones do not talk, and we are left to draw our own conclusions as to why those great beasts no longer inhabit the earth.

In one place the alluvial soil was capped with lava formation. A creek had cut down through the stratum of lava, and also through the alluvial soil underneath, to a solid stratum of bed rock. The alluvium was composed of clay and gravel, and I saw some large bones in it. Altogether I believe that country would be a good field for geological and other scientific researches.

We prospected throughout the mountains, finding a little gold; also some gold and silver bearing quartz. The silver ore was of the base character, carrying lead and copper with the silver. The gold was of the free milling kind, but we attached no value to quartz in that country. At present it is too far away. Yet the time will come when it will be valuable. No doubt good quartz districts will be discovered, for we have seen the quartz, so there is no doubt of its existence.

On going to the summit of those mountains there was another view, and another chain of lakes to the southwest. The country was full of rolling hills, and looked favorable. There was no swamp land, the water from those lakes emptying into a stream to the southwest. We all decided to go to the lakes and make rafts, as we could go around on them on the lakes. Having left our whipsaws hundreds of miles back at our caches, we had no way to make lumber for boats; and the portage being ten miles across, we could not take our boats over. So we packed over our provisions, mining tools, arms and ammunition. We built light rafts and started around the lakes, keeping close in shore, and using poles to shove the rafts along. We prospected the streams as usual, finding a little gold. We had gone down this chain of lakes about one hundred and fifty miles, when we came to the outlet. We could not go far down stream with our rafts; for we would be obliged to walk back again to the lakes. We concluded to leave our light rafts in the lakes, and make two large rafts, — large enough to hold sixteen men each. This would include the whole party. The river being deep and sluggish, we concluded to run down about twenty miles, taking only small supplies with us, as we would be obliged to pack them back. Then to land a party on each side of the river and prospect the small streams coming back, to meet where we left our rafts and surplus supplies. Then to commence the return journey to the mouth of Forty Mile River, where we had cached our winter supplies. Then to build houses for the winter.

Accordingly we built our rafts and ran down about fifteen miles, when the water began to be more rapid and we thought it advisable to stop. Here we saw some signs of Indians, but thought little of it, because the Indians had always been friendly, with the exception of the little misunderstanding I had had with them up the river.

Chapter XXV

YEARS BEFORE I HAD BEEN TOLD by a Russian trader, who had been trading on the Yukon, about a tribe of cannibals that lived on the head waters of the Tananah [Tanana], a river tributary to the Yukon. I knew that the Tananah river was still several hundred miles further westward. Little did we think that we were camped within five miles of those horrible man-eaters. Being at the lowest point we intended to go to, we wanted to stop here several days before starting back. All unsuspicious of danger, we were out prospecting every day. Sometimes we went up small streams, and having mining tools to carry, we did not take any arms with us. Foolish, careless men. It cost us the lives of two of our best men, who were murdered and eaten by those horrible inhuman wretches.

These two men had started in the morning to go to a place where they had been sinking a hole, and having found a small prospect they wanted to try to sink to bed rock. They took nothing with them but their tools and a lunch for their dinner. All came to camp in the evening except those two men. The place where they were prospecting was down the river and a little way back from it, making it about three miles from our camp and directly back about three miles from the camp of the man-eaters.

The next day we commenced searching for them, but without success. We found Indian tracks, and also discovered the Indian camp. No Indians had visited our camp, which satisfied me that they meant mischief. They surely knew that we were there, as we had fired guns at geese and ducks the first day we landed, and had fired more or less every day afterwards.

We were now on our guard. Those who had camped across the river moved over to us. All our arms were put in good order and ammunition got ready. Fortunately all were well armed. All had repeating rifles — principally Winchesters — with two exceptions, and they had breech loading shot guns, which were good with buck-shot at short range. There were also the arms belonging to the two missing men.

No one had been to where the men were prospecting, but we knew the direction they had taken. So in the next day's hunt we found the place. There was no evidence of a struggle there. We hunted all day and found nothing more. Their tools were there and the case was a puzzle. We were now thirty strong. That evening we discussed the matter. We could not think of moving away and leaving the men. It was decided that the following day ten men should, if possible, try to follow the men's trail; that ten should go to the Indian camp on pretense of wanting to trade; that five should stay in camp; that five should go in a direction immediately between the Indian camp and the place where the men had been prospecting; that in case there was any firing, the middle five should at once reinforce the party where the firing was, and that all should then get back to the camp, where we would be governed by circumstances.

We thought it might be that they had made prisoners of our two comrades. I became one of the party who went to their village. We were well armed and determined in our purpose. We walked boldly down the river and without warning walked suddenly into the village. At first it created a great excitement among them. Some disappeared, others, more bold, stood their ground. Having been requested by the other members of my party to act as spokesman, I at once laid down a cloth or piece of calico that we had brought along for the purpose. I placed thereon numerous small articles, such as Indians trade for. This they understood, and at once gathered around in great numbers. I had instructed my men to watch their every movement, and also not to allow them to crowd too closely. I made a line on the sand which indicated to them I did not want them to cross.

They brought forth furs and skins of all kinds, and began to trade. I was not deceived in their appearance. They were an extremely large and powerful race, but few of them being less than six feet in height, and the women were almost as tall as the men. The weather was warm and they wore but little clothing. Some of them did not wear any, being entirely naked. Others wore a skin covering or breech cloth, fastened at the waist, and hanging down almost to the knees. Their hair, which is very black and straight, hung loose down to their shoulders. It was matted and full of

Trading with cannibals.

FROM THE ORIGINAL EDITION

vermin. They were extremely dirty. I do not think they had ever washed themselves. Their features were more like those of wild beasts than those of human beings.Their eyes were of the Mongolian type. Their foreheads were low and receding. In some cases the chin would be receding, in others the reverse. Some of them had heads of unusual length behind, almost amounting to a deformity. All had the wild beast expression of countenance, and showed a lack of intelligence. I have seen a good many wild tribes but have never seen before, and never expect to again, a race of human beings which could give one such feelings of loathing and disgust as these Alaska man-eaters.

Their language consisted of a succession of harsh and unpleasant sounds connected with signs, and I had concluded that they had little or no dealing with traders; as I could not see any articles in the camp which they could have traded for, excepting some old knives and a couple of very old axes. Their arms were bows, arrows and spears. From some source they had got iron that formed their spear heads, although some of them were made of copper. Their arrows were pointed with bone, and some carried clubs with rocks fastened to the end.

I finished trading as quickly as possible, as I wanted to go through the village, to see if I could see any thing pointing to the fate of our missing companions. Their houses were made of sticks, brush, grass and the skins of moose and caribou. We walked boldly through the village, and inspected every thing that came in our way, but found nothing. We then went back of the village into the timber. We saw numerous trails which appeared to lead farther into the timber. I also noticed that there was a great deal of fallen timber close behind the village. I afterwards found I had done a good thing in posting myself so well. We then returned to camp, and had not been there long before the other two parties came in. They had found some evidences of a struggle not far from and farther down the little stream where the boys had been prospecting, but had found no trail leading from the spot. Some of the party wanted to leave, claiming that we could do no more. To this I would not consent, and the majority sided with me.

There was much difference of opinion as to what was the best course to pursue. We decided that the next day a party should go again to the place where the struggle had been and search farther. That night two of our men had gone a short distance from camp in the evening, and were returning late, saw some of the natives sneaking around our camp, but had not been seen themselves. This satisfied me that they were guilty of murdering our companions. I also believed that if we attempted to leave without punishing them they would attack us. The country being rough and full of underbrush, they could kill a great many if not all of us. However, I did not say any thing then, but next morning joined the searching party. We left about one-half the men in camp, the remainder making the searching party. We went to where the struggle had been. The evidence of a struggle was quite plain, and in the creek, close by, we found a button. Evidently it had been torn off there. This satisfied us that they had walked in the water to avoid making any trail. We followed down, and about one mile below found the place where they had climbed out on the bank. There was a great number of tracks, most of them coming from the direction of the cannibals' village. From this on the trail was easily followed. It led us about five miles to a small stream, that ran through a small open glade. Here the horrible tale was told. Several fires had been built, and in the bushes at the edge of the glade, we found the bones of our unfortunate companions. We could see that they been roasted, and some of them plainly showed the marks of teeth.

At once it flashed across my mind that we should keep the man-eaters ignorant of our discovery. I spoke to the others about it. They at once coincided with me. So we did not disturb any thing, but turning toward camp, went back as soon as possible.

Then the question arose, what should be done? A few wanted to leave at once, and I am glad to say only a few. They were the bar-room fighters of whom I have previously spoken. I let them talk and said nothing, until at length several asked me to express my opinion, which I then did.

I told them it was self-evident that the man-eaters were not afraid of us. Had they been afraid, they would have left at once after committing the horrible act. That I believed they were only waiting until we should begin to travel, when they could destroy us with their arrows from ambush without exposing themselves. That I believed they were watching us continually, and since I had carefully reflected on the matter, I believed our precaution about the discovery was useless.

They then asked me what I thought we should do. My proposition was to attack them in their village, kill as many as possible and teach them to respect the whites. We could then return in safety. As we were armed with repeating rifles, and they had no idea of our power, they would be so much surprised that the victory would be easy. Whereas, if we started back without punishing them, we would have the same number to fight and also would do it at a great disadvantage. Having talked the matter over, they adopted my plan and war was decided on. The following morning was appointed for the attack. We decided to start at daybreak, go down through the timber, and attack them from the rear.

The men insisted on my acting as captain of the band, which I consented to do provided all were satisfied, which they were. I then told them that I should expect them to obey promptly any order given by me. To this they consented, and we lay down to rest. I scarcely slept that night, but lay thinking of the possibilities of the coming day. Would we all be alive by the next sunset? I knew that they outnumbered us almost twelve to one, but I had great faith in our Winchesters; besides, fully one-half of the men had Colt's revolvers.

The nights then were very short, there not being over two hours darkness. As soon as dawn appeared I got up, stirred up the fire and got the men up. We made a large kettle of coffee, of which all partook, some also eating a light breakfast. I told the men I wanted them to be very careful and not expose themselves. I pointed out to them how inconvenient it would be if there were many wounded. Although I was positive that we should carry the day, yet we could do it without exposing ourselves. I said that we should keep behind trees; that if they pressed us too hard we should avoid a hand to hand fight by slowly retreating from tree to tree. That when we had them checked we should regain the ground by again advancing from tree to tree. I advised them to make no move which would effect our position, unless I gave the order. I then appointed a man to act as my assistant, my main object being to prevent the men from needlessly exposing themselves. I left five men in the camp, and with twenty-four went through the timber and came out behind the man-eaters' village.

It was as I had expected. They had been watching us and

there was a great clamor and noise. I could see that they were armed and ready. I formed the men in a line with instructions to keep as near in that position as possible, and keep under cover of the trees. Placing my lieutenant at one end of the line, I myself went to the other. This would enable us to watch and if any were wounded we should see it at once; whereas, if the men were scattered too much, we should not know where to look for them, and in case of retreat we might accidentally leave wounded men in their merciless hands.

I then ordered the men to advance, keeping themselves covered as they had done. Closer and closer we got, and the noise increased in the village. I then gave the order to fire whenever they had an opportunity to kill anyone, but to fire no useless shots.

We did not wait long. I saw they were getting ready for a rush toward us. I got upon a fallen tree trunk, but yet managed to keep behind another tree. I saw the natives start, and sung out to the men that they were coming and told them to make every shot count. On they came. The whole horrible, brutish-looking lot came fairly under fire at less than one hundred yards distant. I opened the ball by firing the first shot, and then the rifles rang out on the still morning air, steady and regular. Then I saw what a terrible weapon the Winchester was in the hands of good marksmen.

Down they went by scores, almost every shot appearing to count one, and yet they kept coming. Now they were within one hundred feet of us. They had come to the fallen timber that I have before spoken of. They were obliged to climb up on it to get to us, which exposed their whole bodies at a very short range. In vain. Not one could gain an upright footing upon the timber. When the front one fell the next one tried it, only to fall dead in the arms of his brother cannibals behind. They were checked; they wavered, then broke and ran. They had lost terribly for so short a time. I then ordered the men to advance as far as there were trees to cover their movements. We got so close to their village that we could fire through their frail houses. They fired arrows at us, but we were not exposed and they did us no harm.

A few of them sneaked up behind some fallen timber and used their bows, but it was all guess work. They also threw some spears over a pile of fallen trees, behind which some of our men were. A spear struck one of our men in the knee, making a severe but not dangerous wound. This was the only man of ours who was hurt during the fight. They had had enough of charging us. Taking seven men, I passed down to the lower end of the village where I could keep in the timber and still get closer. There we got a splendid chance, firing on them in front as well as in the rear. They were now getting it too warm to stand it any longer. They became panic stricken.

Some jumped into the river and swam across; others were shot in the water, and some got away through the unguarded space at the upper end of the village. At last there was nothing more to fire at. Some women and children were hid in the huts, but the most of them had got away. When we discovered their hiding places they screamed and ran. We let them run, paying no further attention to them.

We had killed one hundred and four, and about thirty were so badly wounded that they could not get away. Some of our men wanted to kill them, but I objected. I thought we had punished them enough, and I think their appetite for the white man's flesh is spoiled for many years to come. We burned their village and destroyed everything of any value that we could see. We went and got the bones of our murdered companions and buried them, and started on the return trip. We did but little prospecting on the return. All were dejected in spirits. Our provisions were getting short.

We came to our rafts on the lakes and found everything all right. We traveled to the head of the lakes. We crossed the portage to our boats and found the shallow lake on which we had left them dried up. Here was another fix. Our boats were not less than ten miles from water. We had no tools to make other boats, and boats we must have. We decided to carry them. Our provisions were almost exhausted. We had carried our wounded man all the way from the battlefield — if so it could be called when it was so one-sided. Bad as those wretches were, I can never think of them without feelings of compassion. Their punishment was almost too severe. I did not mind the killing of the men. They had made war on us in the most treacherous manner. But if I had it to do again I would not burn their village and their fur clothes and bedding, for that must have caused much suffering among the children in the long, cold winter that followed.

Chapter XXVI

WE CARRIED THE BOATS. It was a hard job. The mosquitoes tormented us continually. There was no game, and our provisions got so low that we put ourselves on rations. Once more we embarked in our boats. There was much shallow water and sand to get through, which delayed us a great deal. We pulled, hauled, tugged and lifted, which increased our appetites, but did not increase our supply of food. Nor did it starve the mosquitoes. Such a life! It was wretched — such hard work with not enough food to sustain us, and the insects blinding us. For several days it remained about the same. We had cooked the last thing we had. It was no use to look for game there. Another day's travel would carry us through swamp lake, as we called it, to open water. No more boating in mud then; but it had to be done without a particle of food. Yet we did it. Now it was open water to the river. Then there was a current to carry us down to our winter supplies. We gained the river and drifted down, only pulling enough for steerage way.

And here is another incident to be recorded; another comrade called to his last account. A small boat that had started first had got fast on a rock in such a manner that they could not get it off. There were only two men in the boat. The rock was near the middle of the stream, in a rapid, dangerous place. While working to get it off, one of the men taking an oar got a position where he thought he could lift it off. He lifted too hard, the oar broke suddenly, and he was precipitated into the whirlpool. The other man on the rock says he never came up, or if he did, he did not see him. The other man stayed on till we came along. We ran so close to him that he could jump into our boat. As we ran down we came to men working on the bars. They gave us food, and we were happy once more. We found our caches all right at the mouth of the river.

Chapter XXVII

AFTER RESTING A FEW DAYS, we commenced looking around for a suitable place to build our winter quarters. Having satisfied ourselves as to the location, we at once commenced building, using logs. We built double walls, making them more than four feet in thickness, and making a roof of logs, bark and gravel, also four feet thick. We filled in the hollow between the walls with clay and gravel. Our fire-place was large and built of rocks. The doors were double, made in such a manner that before opening the inner door we could shut the outer one, thus preventing the cold from getting into the house. There were no windows, for there was no glass to make any. In fact, as there is little daylight in winter in that region, there would be but little use for windows. We built our cabins to accommodate from two to four men each.

The weather was now getting cold, it being the latter part of September. We then went hunting. Knowing where there were some natives of the better class, we went to see them, as we wanted to trade for fur clothing or fresh moose or caribou meat. We got the clothing, but they had no meat to spare. We then went hunting in earnest. Finding a good place, we soon killed game enough to last us through the winter.

As fast as we killed the game we allowed it to freeze, and by keeping it in the shade we kept it frozen. We got our beef to camp and built a small house for it. We then got a supply of wood, and when winter came we were ready for it.

Shorter and shorter grew the days, and colder became the weather. Snow fell to the depth of about one foot, and did not get over two feet deep during the winter. The sun disappeared entirely, but the Aurora Borealis, or Northern Light, came in its stead, making a strong light, not unlike the electric light. The weather became extremely cold, freezing quicksilver, and getting much colder afterward. Not having any way of knowing precisely how cold it was, we could only guess.

When we went out to fetch wood or water we had to dress very warm, even covering our faces and only leaving a hole to breathe and see through. Where the frosty air could come in contact with the bare skin it felt like fire. The little moisture that was in the wood, and what went up the chimney from pots when cooking, would freeze at the top and choke the vent with ice.

When the weather was not too cold we would hunt and trap for rabbits by the light of the Aurora Borealis. The extreme cold was not constant, but the weather moderated at times, so that by bundling up very warm we could stay out several hours. Our cabin was carpeted and lined with the skins of animals. Our bedsteads were made of poles, our table of rough boards and our chairs were blocks of wood sawed the right length. For light we used tallow-dip candles, which we made ourselves from the tallow of the moose and caribou. We also used some bear's oil in a lamp made from a yeast powder can.

I used to go out as much as I could for exercise and fresh air. In the cabin where I lived were two others, there being three of us in the cabin. I induced my comrades to exercise as much as they could in the open air, for I dreaded scurvy, and sure enough it came. The men were not all so well fixed for supplies as I and my messmates were. Some of them were short of provisions, and what they had were not of the best quality. In all there were about forty of us wintering at that place. The scurvy made its appearance about the middle of February. The one attacked would first complain of soreness of the gums, then the teeth would become loose, the legs became swollen and discolored, and if one would squeeze the flesh of the affected part the print would remain where one pressed. We were short of medicine, as many had brought none with them. Humanity demanded that we should divide.

I had taken in among my other supplies some delicacies in the shape of canned goods, milk, jellies, jams, and Liebig's extract of beef. I did not hesitate to produce at once my little stock of delicacies and use them for the benefit of the sick. With medicine I was poorly provided, having only a few boxes of Brandreth's pills, a few bottles of Pain-killer, and a small bottle of gum camphor soaked in alcohol. That was all I had in the medicine line.

We did all in our power for the poor sick men. They

gradually grew worse. We got barks and roots, and made teas, but they did no good. Twelve of our men were now sick and growing worse every day. Two had died. We tried willow bark and that did some good. All used it and it prevented the disease from spreading.

The parts of the body most affected with the disease would turn black, and the teeth would come out. The lower extremities became affected first, the disease gradually working up until the vital parts were reached, when the patient died. The affected parts turned as black as charcoal, and, strange to say, the patient suffered but little pain. Two had already died, and before we discovered the virtues of the willow three others were too far advanced to recover. The other seven recovered.

I was one day sitting talking to one who was near his end. I was trying to cheer him, although I knew he could not live long, yet he did not think he should die. His mind was clear and he was talking of his intentions and plans for the coming summer. He asked me to get him a drink of water. The kettle was empty. I took the kettle and went out to the hole in the ice where we got water from the river. I got the water, and coming in filled a cup and went to his bedside. He did not move. I spoke to him. No answer — he was dead. I had not been gone more than five minutes.

Chapter XXVIII

SPRING WAS NOW AT HAND, although the ice on the river would not break up for some time. We made sleds, and loading them with our supplies, started up Forty Mile River to where we knew of bars with gold enough to pay for working them. We could not work them in the fall on account of high water. The water was now very low. As we found the sledding on the ice very good, we arrived at the bars in due time. We cut and broke away the ice, but found the gravel frozen. We made fires on it and in that manner thawed it out. We worked very hard, as we wanted to take advantage of the low water, for we did not expect it to remain so long. We scarcely stopped long enough for rest and refreshments. We were also finding some very coarse gold, some pieces weighing as much as two ounces. For a while we did well, getting as much as fifty dollars a day to the man. But the gold began to give out and the water began to rise. The ice broke up and went out. We had nothing more to do there. We made rafts and floated down on them to our winter quarters. Here we stayed a few days, resting and planning for the summer. With the gold we found on Forty Mile River we found another metal. It was quite malleable, was in the form of grains similar to gold, and was of a steel-gray color. It proved to be platinum, which is worth about eight dollars per ounce. It was not plentiful.

We had a great many different plans for the summer. Some wanted to ascend the Yukon to prospect; some wanted to remain in the vicinity of where we were, and I wanted to prospect down the Yukon as far as I could find any mountains to prospect, and in the fall go out at the mouth of the river and go to St. Michaels, a trading post on Bherings [Bering] Strait about ninety miles north from the mouth of the Yukon, belonging to the Northwest Alaska Commercial Company. There we should surely be able to meet with a whaling or trading vessel and get transportation to San Francisco or some other southern port.

Chapter XXIX

SIX MEN AGREED TO GO WITH ME and we at once made preparations, bid good-bye to our friends, and started on our long journey. We drifted leisurely down with the current, stopping and prospecting at every favorable looking tributary stream as usual, and finding a little gold. I also noticed considerable quartz, and I have no hesitation in saying that the time will come when the interior of Alaska will be a great producer of the precious metals. At times, if we found gold at the mouth of a stream, we would ascend it, sometimes a considerable distance, with always the same result — nothing to pay. And so the days, weeks and months came and went and autumn had come. The valley had increased in width until we could no longer see the mountains on either side except at times, and then only in the hazy dim distance.

So we decided to give up prospecting and go to St. Michaels. The river was now very sluggish and filled with countless islands. It was also very wide, in some places fifteen or twenty miles. We had passed the mouth of the Porcupine River, which comes in from the north and the mouth of the Tananah [Tanana], a large tributary coming in from the south, near which a great many Indians were camped. We stopped at their village, and here was a sight to look at. The filth and dirt was indescribable. Their poverty was almost beyond belief. Their food consisted principally of salmon, with some geese and ducks. The fish, geese and ducks were dried for winter use. There appeared to be no other kind of game. As they had no skins nor furs, their clothing was made from the skins of the salmon sewed together.

Their houses were dug out of the high sandy river banks and covered with poles, with a smoke hole in the top. The front of the house was built up with poles. They were quite friendly. I went inside of one of those houses, or dens, to satisfy my curiosity, and oh, the smell! I cannot see how a human being could live in such an atmosphere. My curiosity was fully satisfied. I never wanted to examine another.

We passed on down to the mouth of the river, in many places passing small camps of those wretched people. They appeared to be more like esquimaux than the up-river natives, and this feature in their appearance became more and more marked as we went down the river.

Now the country became entirely destitute of timber. The banks were low and the river was divided up into many channels. The surrounding country was very low and flat and covered with a brownish lichen and moss. We were now down to where the tide from the sea backed the water up. Not a tree nor bush to be seen in any direction. But we could get plenty of driftwood that had come down the river, a fortunate thing for us, for the nights were quite chilly. Ducks and geese were here by millions, and we were well supplied with food as we still had plenty of other supplies. But what a land of desolation! And yet we would occasionally pass a few rude huts made of driftwood.

Well, as all things have an end, so did this long river, and we found ourselves once more on the shores of the briny ocean. Our boats were not fit for ocean traveling, so we got some logs of driftwood large enough to make lumber. We broke up our two river boats, and sawing some lumber from the logs, we got enough to make one sea boat large enough to carry the whole party to St. Michaels, which was about ninety miles distant.

Chapter XXX

SO ONE BRIGHT MORNING WE STARTED, but the weather did not remain good. The sky became overcast and the wind commenced blowing; however, the wind was not strong, so we kept on. Our boat proved to be a good sea boat, riding the seas as bouyantly as a duck. The days were now about twelve hours long. The nights were very dark, there being no moon. We decided to camp for the night at the first favorable place, but we could not find the favorable place. We wanted a place where our boat would be sheltered from the wind and waves, as the coast was rocky, but we could not see any place we could haul the boat clear of the surf. Daylight disappeared and we had found no landing place.

The wind increased and blew a perfect gale from the southeast. We were obliged to keep away from the shore, for there was danger of striking rocks. Then along with the wind came a cold, blinding sleet. Our condition was one of great discomfort. It was so dark that we could not see each other in the boat. We had had no supper; were cold, hungry, and wet to the skin, with a prospect of passing the whole night at sea. The wind was driving us rapidly to the northwest. I knew that the Straits here were not very wide, for during the day we had got a sight of what I took to be the Siberian coast. The wind increased. It was now almost a hurricane, but the rain and sleet ceased. All we could do was to pull with the wind for steerage way. The boat could not be handled properly, for the seas were heavy and breaking and we shipped a good deal of water.

Two of the men were experienced seamen, which I believe saved our lives on this occasion. They handled the boat well, and yet we shipped so much water we had to lighten the load. First we threw overboard our mining tools; next we threw over several sacks of flour, and yet we had to bail continually. Our rolls of bedding came in the way of bailing, so over they went. Nothing was left now but a couple of sacks of flour and a few other kinds of provisions, and they being saturated with salt water, would be but little better than nothing.

However, it was not destined that we should keep even these. The wind continued on in its fury, and we took turns bailing. Now we thought we could hear the roaring of the surf on the beach ahead. It was the shore of an island near the main land and the hurricane was driving us directly onto it. To hold the boat off by pulling at the oars would be impossible against such a hurricane. As we came nearer we could see the breakers. I knew the struggle was near. If the undertow was not too strong it was quite likely we could make the shore. On the contrary, if the undertow was strong, to gain the shore was almost impossible. Day was now breaking, which was a good thing for us. We tied our rifles securely inside the boat, also our revolvers and cartridge belts. We lightened ourselves as much as possible. It was now lighter and we could plainly see timber on the beach.

The experienced men proposed to send the boat ashore under all the headway we could possibly give her. Then, if the beach proved to be soft, the headway would carry us so much further up. So we headed directly for the shore and pulled with all our strength. When the shock came it was a heavy one. I heard the timbers breaking. We had struck rock bottom. Another wave struck and swung the boat around, and then it was lifted and came down and struck on the side. We were all thrown into the water. All hands made for the shore. Another wave came and carried me off my feet and then receded. I managed to get a footing and did not go back with it. I again struggled for the beach with all my might, and before another swell came I was out of danger. The others were as fortunate as I, and we were all safe so far as drowning was concerned.

Chapter XXXI

WE STOOD ON THE SHORE and watched the boat, for it was now light enough to see a little. I was very anxious to get my rifle. It was tied in the boat, and the boat was already a wreck. To get our arms was impossible. The boat went entirely to pieces. We had lost every thing — our boat, our tools, our beds, provisions, and even our arms. We were in a wretched plight. Not a match, nor anything to make a fire with; we had only the clothes we wore. We were landed, we knew not how far from St. Michaels — the nearest place that we had any knowledge of where we could get assistance. The worst loss was the boat.

As soon as it was light enough we commenced to explore the island, for an island it proved to be. There were no signs of natives, although I saw plenty of deer tracks along the beach. However we had no way to kill them and no fire to cook them. We went around the upper part of the island. It was not more than three miles to the main land, the coast of Alaska further northwest.

I could not see any smoke or sign of inhabitants as I viewed the coast of the main land. We gathered clams and mussels and ate them raw, for we could not cook without a fire; but they kept us from starving to death at least. We thought it best to stay at the upper end of the island so we could watch both sides for natives who might be passing in canoes. It would be our only chance to get away.

It rained almost the whole day. We were very wet, cold, hungry and miserable. Now night came on. I can hardly tell how we lived. We tried to get a fire by rubbing two sticks together, but we did not succeed.

The next day the sun came out and dried our clothing. Fortunately for us the sun shone several days in succession. We gathered fallen timber and made a kind of a house. We roofed it with brush and such barks as we could find loose. Then we gathered leaves and moss and made a nest, and found it much better than no place. The clams and mussels were hard living, and especially so as they were raw.

We tried very hard to get a fire. We rubbed sticks together until they would smoke, but we could not make them burn. We had heard of savages making fire in that way, but it would not work for us. We then got rocks and by striking them with our pocket knives we could get sparks, but could not get anything dry enough to catch from them. Still we kept on trying, and no doubt would have eventually succeeded. But one day two canoes of natives, apparently a hunting party, came close to the island. We managed to attract their attention. They landed and were very much surprised to see us. They had fish and venison and also matches. They were friendly and kind to us. They camped on the island two days and hunted deer. They gave us fish and venison. We made large fires and had large roasts. We had lived on raw shell-fish and been without fire ten days. I thought the venison and fish were the finest I had ever eaten. I even felt as if the fire was of an extra quality. In this manner will one appreciate the necessaries of life if for a time they are deprived of them.

I then tried to hire them to take us to St. Michaels. They agreed to go, but wanted to know in what way and how much we would pay them. I showed them money — silver coin. They took it into their hands and fitted it to their clothing, which intimated that they only considered it good for buttons. They further intimated that they did not want silver. I showed them gold coin and gold dust with the same result. They would not take money, but indicated that they wanted our clothing.

After a good deal of dickering I made them understand that we could not spare our clothing, but if they would take us to St. Michaels I would get anything for them that they wanted that was there for sale. To this they agreed, and the next day we started. Those natives were clothed in furs and buckskin. They lived at some place on the Siberian coast.

From the island on which we were wrecked I could see far to the northwest into Alaska. The landscape sloped gradually back as far as the eye could reach. There were no signs of inhabitants on the land, that I could see. Near the coast it was covered with scrubby timber. Further back it had the appearance of a prairie. When the weather was good, the air was free from smoke, the sign that always indicates the presence of cities or a settled country. From the summit of a large hill I could see in the north and west the dim outline of a range of mountains. There was also a vast, level plain visible to the view.

We now embarked for St. Michaels and on the evening of the second day we reached it. It is a trading post. The people were kind and hospitable and furnished us with a house to live in. The post is on a small island, quite near the main land. The water is bad, and not a stick of timber grows within many miles. There is plenty of driftwood, however, to be picked up on the beach. The trading company make this their supply station for all their most northern trade.

Here we found an interpreter and made satisfactory arrangements with our Indians. We allowed them to name the amount that they thought we should pay them. Their demand was very light, not amounting to more than one dollar to each man of the company. They expressed themselves well pleased with the payment. We then told them through the interpreter that they had treated us well and now we would make them a present. We gave each one a good, heavy pair of blankets. At this they were very much surprised, as well as highly pleased. They promised us in return that they would always be kind to and friendly with the whites.

We remained at St. Michaels nearly three weeks when a government revenue cutter stopped. The captain gave us a passage to Sitka, and my travels in interior of Alaska had ended.

I stayed at the coast several years, sometimes prospecting, at others in the employ of the different mining companies. I do not think that I shall ever again return to Alaska, although I consider it a good field for the prospector. My constitution is broken and I can no longer endure the hardships attending such a life. Through all those long years there was an overruling Providence; an ever helping hand, whose presence was never wanting; who, through the medium of our conscience, pointed out to us the path of duty. Its power is felt by all right minded men and women, and if we obey that voice it will not only bring us happiness in this world, but also in the time to come. Who is it that having done a good act, an act of kindness or charity, that does not feel happier when he reflects on it? And are there any, who having done a mean action, do not feel degraded in their own estimation? Such is the gift of conscience that the All-Ruling Power has given to us, that we may be able to know the right from the wrong.

I deem it proper here in conclusion to add a statement that of the men who were killed or died on those trips I only know the names of two. They are Thomas Jones, drowned; and — Jackson, murdered and eaten by cannibals. Usually among miners only nicknames are used. Thus it is seldom you know the name of any, except a very intimate friend.

In speaking of matters pertaining to the education of the native Alaskan, it is only fair to mention that the Society of Friends is represented there. Their headquarters are at Douglas City, on Douglas Island, where they have built a school house and made other improvements under the able management of Mr. E.W. Weesner. Like the Pennsylvania Quakers in William Penn's time, they are rapidly gaining the good-will and confidence of the natives.

These missionaries are exceedingly well liked, not only by the natives, but by all who have dealings with them. I believe that much good will result from their exertions, and this may be said of all our missionaries who are laboring in Alaska. They are zealous and untiring in their efforts, though they have had many difficulties to contend with. A close observer can plainly see that their good works are commencing to tell for good, and I believe it to be the duty of all who can to contribute to their support, no matter what denomination they may belong to; for a number of them have spent many years of the best part of their lives in this good work.

Epilogue

On or about the first of October, 1886, Walter Pierce got a job running the Music Hall at what was then usually known as Newtown but which soon became the town of Douglas, on Douglas Island just north of the Treadwell mines. The place consisted of a sizeable community of Indians from all parts of Southeastern Alaska, many of whom worked at Treadwell; a number of saloons and dance halls, some family dwellings, a good many cabins and shacks occupied by single men, and one or two small general stores. For the most part, the white miners and others who worked at the mines lived at Treadwell, in company housing.

Most of the saloons in Douglas at that time were, from all reports, pretty low dives and the Music Hall was probably no exception. The name implies that in addition to a bar it had a dance floor for what were locally known as "squaw dances," since Indian women were the only female partners available to most of the miners. Selling or giving liquor to Indians was illegal at that time, but the law did not do much of a job of stopping either sale or gifts. Years ago a Douglas old-timer outlined to this writer one foolproof method used in the dance halls:

"You see," he said, "when you took an Indian girl to a dance, she wanted a drink, just like everyone else. But there was usually a deputy marshal hanging around and we had to be careful. There wasn't any law against kissing a squaw, so what you did was go to the bar and buy two shots of whiskey. One of them you drank — and swallowed. The other one you drank but didn't swallow. Instead you went over to your partner and transferred it from your mouth to hers."

Exactly what happened in the Music Hall late in the evening of Saturday, October 16, 1886, or early the following morning is not entirely clear, but the nearest newspaper, *The Alaskan* at Sitka, called it a "riot." There were a good many drunken Indians around, and perhaps an equal number of drunken whites. The Indians demanded liquor from Pierce and he refused them, and when they grew insistent he ejected them from the hall. He sustained some cuts and bruises around his head and face in the process. The Indians began smashing windows in the hall and two pistol shots were fired. When the smoke cleared an Indian from Kake, Klu-keetz, was dead, shot through the heart, and Frank Mehrwalt, the captain of the Alaska Mill & Mining Company's boat, the *Lucy*, was wounded in the chest.

Both men were innocent bystanders. Klu-keetz worked at the mine and was said to have been quiet and industrious and not to have been drinking. Mehrwalt had been sitting in Hawthorne's Saloon, several doors down the street from the Music Hall, when he heard the sounds of the riot. He stepped out the door just as the shot was fired. A friend of his, George Foster, ran to get Dr. Wyman, the Treadwell mine physician, but was set upon by some of the rioters and it might have gone hard with him had not a group of Haida Indians been passing by and rescued him. Mehrwalt's wound was not serious and he recovered. Pierce got a boat and pulled over to Juneau and got Deputy Marshal Phil Weittenhiller, who arrested several men, including Pierce.

Pierce was arraigned on a charge of murder and was indicted by a grand jury at Sitka in the May, 1887, term of the U.S. District Court. Willoughby Clark, who had been in the country since Cassiar days, was his attorney and asked for bail. This was denied and Pierce remained in jail until the fall term, when there were two murder cases. The other was Frank Fuller, charged with the murder of Bishop John Charles Seghers on the Yukon

River. Pierce went to trial first, on Tuesday, November 22, 1887, before Judge Lafayette Dawson. There were 87 veniremen and Clark challenged 20 of them and the government challenged 5 before a jury was selected.

The testimony was conflicting. Several Indians asserted that they saw Pierce lean out of a window of the Music Hall and fire the shots. A white man said that Pierce had so much blood on his face from his wounds that he could not have aimed a pistol. It was pretty clearly established that Klu-keetz was at least sixty yards from the nearest part of the Music Hall when he was hit and Mehrwalt was even farther away. The jury was out 23 minutes, returning at 3:55 p.m. with a verdict of "Not Guilty."

The next report we have on Pierce is a small item in *The Juneau City Mining Record* for June 6, 1889: "Walter Pierce, who has been quite feeble for some time, expresses thanks to his friends on Douglas Island who raised a purse for him before his departure on the *Ancon.* The purse amounted to more than $500 and was raised in both Juneau and Douglas." And in August, 1889, John Olds of Juneau received a letter from Pierce saying that he was running a fruit and cigar store in Denver and that his health was much improved.

Pierce's book was published in Lawrence, Kansas, in 1890. On the flyleaf of the copy in the Alaska Historical Library at Juneau is the name "John T. Moore" and in what seems to be the same handwriting: "This book was bought of W.H. Pierce 1890. He was far gone with consumption at this time and used large quantities of morphine to ease pain. He died within a year (if not within *the* year). The book was written to help support himself."

That was only thirteen years after he first set foot in the North as a young man. It seems probable that he was not older than 40 when he died, and he may have been considerably younger. It was not unusual for the miners and prospectors of the frontier North to die young. What was unusual was for one of them to leave us a record of his travels in Alaska. Walter Pierce was an exception and we are grateful to his memory.

—R. N. De Armond